Engineer Your Software!

Synthesis Lectures on Algorithms and Software in Engineering

Editor
Andreas Spanias, *Arizona State University*

Advances in Modern Blind Signal Separation Algorithms: Theory and Applications
Kostas Kokkinakis and Philipos C. Loizou
2010

Advances in Waveform-Agile Sensing for Tracking
Sandeep Prasad Sira, Antonia Papandreou-Suppappola, and Darryl Morrell
2008

Despeckle Filtering Algorithms and Software for Ultrasound Imaging
Christos P. Loizou and Constantinos S. Pattichis
2008

Engineer Your Software!

Scott A. Whitmire

ISBN: 978-3-031-00402-5   paperback
ISBN: 978-3-031-01530-4   ebook
ISBN: 978-3-031-00018-8   hardcover

DOI 10.1007/978-3-031-01530-4

A Publication in the Springer series
SYNTHESIS LECTURES ON ALGORITHMS AND SOFTWARE IN ENGINEERING

Lecture #21
Series Editor: Andreas Spanias, *Arizona State University*
Series ISSN
Print 1938-1727   Electronic 1938-1735

MATLAB is a registered trademark of The MathWorks, Inc., 3 Apple Hill Rd, Natick, MA.

Engineer Your Software!

Scott A. Whitmire
Mayo Clinic

SYNTHESIS LECTURES ON ALGORITHMS AND SOFTWARE IN ENGINEERING #21

ABSTRACT

Software development is hard, but creating good software is even harder, especially if your main job is something other than developing software. *Engineer Your Software!* opens the world of software engineering, weaving engineering techniques and measurement into software development activities. Focusing on architecture and design, *Engineer Your Software!* claims that no matter how you write software, design and engineering matter and can be applied at any point in the process. *Engineer Your Software!* provides advice, patterns, design criteria, measures, and techniques that will help you get it right the first time. *Engineer Your Software!* also provides solutions to many vexing issues that developers run into time and time again. Developed over 40 years of creating large software applications, these lessons are sprinkled with real-world examples from actual software projects. Along the way, the author describes common design principles and design patterns that can make life a lot easier for anyone tasked with writing anything from a simple script to the largest enterprise-scale systems.

KEYWORDS

software engineering, design, design patterns, architecture, problem solving, software quality

For:

Diana Bell, my wife, for her unwavering encouragement.

*To my son Trevor, the engineer among my kids,
who hadn't been born when I wrote my first book
and understandably felt left out.*

Contents

Preface

I craft systems out of software to solve problems and have always considered myself a toolmaker. I build things others use to make their lives easier. The problem is paramount; the software a means to an end. I haven't always used software, but the days of a solution requiring no software are limited, if not already gone. I've been doing this for nearly 40 years, and the thrill of seeing a system fire up for the first time never gets old.

Some people write software for a living; others because they have to. Both groups can enjoy that thrill more often and with less frustration by applying a little thing called software engineering. Software engineering isn't about writing code; it's about building systems out of software components that solve a problem reliably and efficiently. Anybody can write code, but it takes a fair bit of learning and experience to build systems that can range from a set of short functions in a single source file to enterprise-wide applications composed of hundreds or thousands of source files. The only difference is scope. Sure, wider scope adds complexity, but narrow scope doesn't mean easy as even small systems have components that interact in ways that can cause trouble if you don't deliberately engineer them.

Engineering software is hard, and has only gotten harder. Gone are the days when you could write a single program that does anything useful without relying on various external components and functions that already exist. We have a name for writing those single function programs now; it's called scripting. You write a script that invokes many other tools, all external to the program you're writing.

Systems today, especially useful systems, are complex structures of interacting components. Large, complex structures of any kind, physical or abstract, cannot be just built. They have to be designed, and as complexity grows, they have to be engineered. The main goal of engineering as an activity is to prevent failure (of the structure, the success or failure of the engineer is a different problem).

It is ironic that software, one of the most complex creative activities undertaken by humans, is the least engineered. This is due, in part, to history. When a physical structure failed, people got hurt, so we figured out how to analyze designs to prevent failures. The history of iron railroad bridges in England during the 1800s is a good example of the discovery of the need for and development of engineering. Failures in early software weren't harmful so much as annoying, until we started to rely on it to fly our airplanes, drive our cars, run our power plants, automate our companies, and even our homes and personal lives. As software systems have become larger and more complex, the consequences of failure have become more severe. Today, software needs to be engineered, all of it.

Over the course of my career, I have designed and built many large applications, sometimes as part of a team, sometimes on my own. Along the way, I have learned many lessons unique to the world of large systems. We don't often build large systems from scratch anymore, crafting them instead from components built by others. This new way of building systems is actually more difficult, but even then, we always end up writing some new code.

All of the software I've written was for internal consumption of an employer or client. Some of that was subsequently sold to others, but I have not worked in the commercial software market. This declaration is part disclaimer and part setting of context. There are valid arguments that developing software for the commercial off-the-shelf market is different than for internal use, but most of the real differences are minor. For example, commercial software dictates the operational context (in the form of "system requirements") while internal software is built for a specific, pre-existing operational context. In the end, software is software, and once we know the context in which it operates and what it needs to do, we are left with basic software development.

I currently work in medical research creating the tools used by a cancer research lab to manage and analyze clinical and imaging data to understand the mechanics and behaviors of tumors. Most software in research is created by people who have a different job, and whose training in software development might be limited to a few courses to learn just enough programming to build their analysis tools. For these people, software is a means to an end, and rightly so, but that focus often shows in the quality of the software. This isn't to say that software built by researchers doesn't work as it often works quite well, but it can be very hard to use, even harder to reuse, and is extremely fragile.

Creating software that works is a craft, just as all engineering work is a craft. There are tools and techniques that help you find reasons your design might fail, but none of those tools, save the experience of yourself and others, tell you where to start. I aim to help fix that problem.

Sit back, relax, and enjoy the ride. Maybe you'll learn something. I still do, even after all these years.

Scott A. Whitmire
Scottsdale, Arizona, 2021

Who Should Read This Book

I mentor and work with people who must write software as part of their job, but whose job is not to write software. It takes a lot of training and (mostly) experience to learn how to engineer good software. Through this book, I hope to make that journey just a bit easier. I want to communicate that engineering software that works and works well is not as difficult as it seems, though it is far from easy. Through continued practice, I have distilled the many lessons learned into the material presented in this book.

This is not a cookbook; I don't present recipes for specific problems. The solutions I present solve the problems I describe, but the *types* of solutions are good for any similar problem. This emphasis on types and *shapes* is deliberate and important. Rather than give you the answers, I want to show you how to look at a design problem and create a solution that works in your situation. Feel free to tweak these solutions as required to fit your problem. Learning to recognize the basic pattern of a problem is the first step to becoming a successful engineer. Learning how to modify a pattern to fit a problem better is the second. You will get a lot of both in the material that follows.

How This Book is Built

The book is designed to take you from the beginning of a project to just before you start writing code. The sequence of topics does not imply that you design the entire system before you write any code, but there are activities that need to be completed before code gets written.

Chapter 1 describes the goals and techniques of software engineering in a general sense. It emphasizes that engineering is a verb, not a job title, so you *do* software engineering, even if you *aren't* one. The chapter presents several design principles that will keep you out of trouble, and design criteria that allow you to make decisions along the way. With this information, you can determine that one design option is better than others, and more importantly, understand why.

Chapter 2 focuses on architecture. This will be the shortest essay on software architecture you will ever see and examines the practical aspects of recognizing the shape of your problem and letting that determine the shape of your solution. You will learn that non-functional requirements, those that describe the world in which your software will operate, have far more influence on your architecture than functional requirements. The chapter closes with choosing your programming style and design approach, which, despite appearances, are indeed part of your architecture.

Chapter 3 is about correctly describing the problem you are trying to solve. The skills you need most here are those that you used when solving algebra story problems in high school. Your non-functional and functional requirements are a mishmash of things you need to do, things you cannot do, and constraints about how you do all of them. Translating that into an initial model you can use to build good software is one skill that most programmers never learn. To my continued surprise, I excel at it, and I try to pass on some of the tricks and techniques I've picked up over the years.

In Chapter 4, we actually design software. Software design is about translating the problem description into a set of components that will reliably and efficiently solve the problem. Engineering comes in when you make choices among ways to solve that problem, and when you analyze your end product to be sure it will work as you expect. Engineering has been described as imposing the scientific method onto the design process, an apt analogy.

Chapter 5 considers the world in which your software will live. It's a mean, hostile place, and your design needs to be able to survive. Your software, if it's any good, will inevitably change over time, and you need to be able to control the scope of those changes and easily deploy them when necessary. After introducing a couple of principles that could very well save your life, the chapter describes a series of rocks under which you will find monsters and bugs that will bite if you're not careful. Some of these rocks are well known; others I've learned about the hard way. In both cases, you will learn what you can do to keep the monsters at bay.

Scott A. Whitmire
June 2021

Acknowledgments

There are always a lot more people involved in a work like this than you might imagine and I name but a few here. To all of these and others: Thank you. Thank you for the help you've provided, for the arguments we've had that polished both our views, and for the experience. Mostly the experience. The last 40 years have been a great run, even as it continues.

Margaret Eldridge was the copy editor for my first book and suggested I write this one, so this is all her fault. I owe you a debt, Margaret.

Kamala Clark-Swanson was a colleague-then-boss on the project I use for the case study. Together we made it work, literally. Kamala and I, along with a few other colleagues on that project, solved every one of the real problems I describe in Chapter 4.

Robert Martin, yeah, that Robert Martin. If you Google my name, you still see some of the arguments Robert and I had on Usenet and several list servers dating back to 1990 while he was developing his set of design principles. Those arguments, among others, shaped my understanding of software as a thing, and apparently his as well.

The early gurus of software engineering—Tom DeMarco, Larry Constantine, Grady Booch, Meiler Page-Jones, and others—are real people eager to help, teach, or just talk with those who aspire to learn the craft. Over the course of my career, I've met all of them more than once and can point to specific conversations in which I've learned something profound. This book is but one way for me to pay forward the benefit I've gained from those who've gone before.

The students and co-workers I've mentored over the years helped me understand what was not being taught in Computer Science and Bioinformatics programs and what was needed to help them create software they could write, use, and continue to focus on research rather than tweak code.

Scott A. Whitmire
June 2021

CHAPTER 1

Engineering is a Verb

Most of the time, all you know is where you are and where you want to be. The rest is the adventure.

Writing software is hard; writing good software is very hard, especially when it isn't your primary job. I currently work in a research lab and see many highly trained scientists try to write code to support their research. To be blunt, much of the resulting software is not good. If it works at all, it is hard to use and fragile, and forget about trying to reuse it. Even well-known and well-used tools created by labs funded to build software tools suffer. (Ever try to install dcm4chee, a tool to manage medical images?)

Edgar Dykstra said writing software was more difficult than theoretical physics [1]. Small wonder, all software work is done in the abstract: we conceive of designs, write code, and execute code, but we see only the indirect results of that execution, never the actual execution.

Jack Mostow [2] suggests that the purpose of design is to construct a solution that:

- "Satisfies a given purpose

- Conforms to limitations of the target medium

- Meets implicit or explicit requirements on performance and resource usage

- Satisfies implicit or explicit design criteria on the form of the artifact

- Satisfies restrictions on the design process itself, such as its length or cost, or the tools available for doing the design."

Since these goals are very often in conflict, it's clear that the designer is faced with an optimization problem in which no option is clearly better than any other. These tradeoffs force the designer to choose between various options at multiple points along the way. How those choices get made is the difference between a designer and an engineer (who is also a designer).

My aim is to make creating better code just a bit easier, and certainly more reliable. By intentionally thinking through a short design process, mixed with just a little engineering, you can create better code the first time even if it isn't your main job. I will show you that design and engineering is not as much work as the usual code-test-recode (and repeat) cycle we've been using for the last six decades.

Every new software tool or technique promises to eliminate the need for design. Just jump in and start writing code. Trouble is, design happens whether you want it to or not. If you've ever written code, you've no doubt found yourself at the end of a chain of if-then-else statements and thinking: "Well, now what?" This is design.

Software has become more complex because we ask it to do more and because the environment in which it runs has become more complex. Deliberate design is no longer a luxury; it is essential, but insufficient. Some basic engineering helps make better design decisions. Engineering adds rigor to your design thinking, helping you make decisions based on the data at hand. You encounter a lot of decisions during the creation of software, often involving options with no obvious choice. One option is always better than the others, and engineering helps you find it.

1.1   WHAT IS ENGINEERING, EXACTLY?

Engineering is a verb. The whole purpose of engineering is to avoid failure [3]. Engineering is a mental activity that is part creative and part critical. The creative part considers a problem and conceives of a solution. The critical part, which requires a lot of creativity, looks for ways the conceived solution might fail in an effort to determine whether the solution is safe, sufficient, and economical, in that order.

The engineer looks for ways the product can fail and designs around them before the product is built, stopping only when no other modes of failure can be imagined. Software people don't think this way: we build the first design that comes to mind, or worse, start writing code before we even have a design in mind. Trouble is, we no longer have time to build a design then rebuild it once or twice more after we discover it doesn't fulfill even basic requirements. Engineering doesn't take as long as it sounds, and certainly takes less time than building something twice, or three times [4].

Most programmers behave as if the purpose of design is to find the first solution to a problem that works. Engineering goes further: it's about finding the best solution to the problem given the goals and constraints of the project. The source of these goals or constraints can be the problem itself or may be imposed by the client or sponsor of the work. More often than seems reasonable, these goals and constraints are in conflict, with no way to happily achieve all of them. Engineering, then, is an approach to Mostow's optimization problem in which you navigate a series of tradeoff decisions, looking for the best option given your particular set of goals and constraints [5]. Different goals or constraints might lead you to a different conclusion for the same problem.

As a way to get us all on the same page and to bring this section to a close, I include here the definition of software engineering first published in [5]:

Software engineering is the science and art of designing and building, with economy and elegance, software systems and applications so they can safely fill the uses to which they may be subjected.

This definition, borrowed from the British Society of Structural Engineers, carries a lot of freight. Let's unpack it a bit.

The definition makes it pretty clear that software engineering is both art and science; art applied to a practical purpose. Conceiving a design requires all of the skills of an artist; analyzing the design requires the knowledge and rigor of a scientist.

The definition covers both the design and construction of software. Much of the engineering done on software happens elbow-deep in the code, when you encounter a detail you hadn't foreseen. You bring your best practices and tools to bear on every project, unless there is a compelling reason to use something else. Further, the definition includes maintenance, admonishing you to exercise the same care when modifying the software used when creating it.

The meaning of "economy and elegance" is truly in the eye of the beholder, varies with each project, and recognizes both practical and artistic needs. Economy dictates that the engineer use just enough rigor to solve the problem safely and correctly, but no more. Some problems just don't justify the extra work, others demand it. Elegance implores the engineer to keep the solution as simple as possible. Some level of complexity is inherent in the problem, the choice of tools and process will add more, and the design itself will add even more. It is up to the engineer to add as little complexity as possible while still solving the problem.

The last phrase, "...so they can safely fill the uses to which they may be subjected" is the most important. It covers both what the software *should* do and what it should *not* do. You build systems to solve problems; you have in mind a certain way of working when you build your solutions. Yet anyone who has ever used a screwdriver as a pry bar or chisel knows that all tools will eventually be used in a way that was not intended. The notion of safety has multiple parts. First, do no harm. Operation of the system should not damage anything, including the reputation of the engineer or its users. Second, the software should fulfill its intended purpose without failure, or should fail gracefully. Finally, the software should withstand unintended uses, or safely resist such use.

That simple definition puts a lot of burden on the software engineer, and rightly so. Today's software solutions are relied upon by dozens, hundreds, thousands, or even millions of people. Depending on the application, lives may depend on it.

1.2   ENGINEERING AS EXPERIMENT

If you think about it, designing and writing a piece of software is an experiment. If you build the code just so and execute it in the right environment, you expect to see certain results. When you don't see what you expect, you troubleshoot your experiment, modifying it until you finally see those expected results.

Before you can design and build your experiment, you need some understanding of what you expect to see and why. As your experiment gets more difficult, you have to spend more time understanding your expectations as well as thinking about how you plan to go about meeting them.

Most experiments start with an observation and work to figure out why the observed phenomenon occurs or the mechanisms behind it. With software, we start with a set of expectations and work to figure out how to create them. The ends are different, but the means are the same.

Every experiment starts with an hypothesis, some explanation about the observation and what lies behind it. The hypothesis in software, and in all engineering according to Petroski [3], is the design itself. A design states either implicitly or explicitly that the structure built according to the design, using the specified components, and used within specified guidelines will fill the specified function for the specified period of time without fail. Like any hypothesis, a design can be proven wrong by a single exception to the intended behavior.

Engineering is thus a succession of hypotheses, each built upon the failure of all prior hypotheses. Each iteration is tested or analyzed in an effort to understand how the design will behave under real or imagined conditions. Calculations may be required, which requires the use of measurement. If the new design fails, the failures are found, components redesigned, and the analysis is performed again.

The process is done, according to Michael Nygard [6, p. 6], when the engineer is satisfied that the design is good enough for the current expected stresses, and they know which components will need to be replaced as those expected stresses change. Yes, the definition of "done" is rather subjective.

Design decisions often boil down to choosing between alternative patterns. Different patterns have different characteristics, and you use engineering analysis to evaluate and select among different design options.

1.3   SOFTWARE SHOULD BE CYNICAL

Nygard says that enterprise software must be cynical [6, p. 23]. He's wrong; *all* software should be cynical. The world in which software runs is an inhospitable place populated by users who just want to get a job done, system administrators intent on keeping everything up to date, operating environments that seemingly change on their own, and you, its creator, who wants to continually change things. All of these forces conspire against your software. It can't trust anything in such a world, not even itself. To survive, your software has to resist becoming too intimate with other systems, and put up internal barriers to protect itself from failures.

Software should just *work* first and foremost, and keep working no matter what. After that, you can make it elegant and fun to use.

1.4   ENGINEERING IS *NOT* "ANTI-AGILE"

Agile development has been mistakenly viewed as a method for *developing* software. A careful read of the Agile material reveals that Agile is actually a method for *managing* software development. Agile is not anti-design, but rather "anti-stop-everything-and-do-all-the-design-now." It's about managing the design process, not actually how to do design.

Regardless of when you do it, design happens. Design decisions get made throughout a project, no matter how you organize it. Agile does not eliminate design, but moves it into the code writing activities. Engineering as described in this book is a way to make those design decisions with just a bit more rigor, and with a higher likelihood of success.

1.5   TOWARD MORE PRINCIPLED DESIGN

Design principles help guide your thinking as you work through the process. Think of them as the lane markers on a road; they don't prevent you from doing something stupid, but they let you know when it's about to happen. Most of the design principles in this section have been kicked around by Robert Martin and others, including me, since the late 1980s. Martin has published them a number of times, but the most accessible descriptions for the first five are in [1]. Martin's principles are all written for classes in object-oriented programming, but apply equally well to other forms of software. I added the last two as a result of a root cause analysis of some design disaster, the software equivalent of an aircraft crash investigation.

1.5.1   SINGLE RESPONSIBILITY

Robert Martin first described the Single Responsibility Principle [1, pp. 96–103] in 2005 as follows.

> *A class should have a single responsibility, that is, only changes to one part of the software's specification should be able to affect the specification of the class.*

The example he gave then was separating the gathering of content from the formatting or rendering of that content (Figure 1.1).

The principle given in [1] is different; he changed the definition in 2014. I prefer the original. Following the original version will lead to highly cohesive abstractions, whether they be classes or other structures in your code. You can prove this out by comparing the cohesion of a class designed according to this principle to one that is designed not following it using the measure defined in Section 1.6.3.

1.5.2   OPEN-CLOSED

The Open-Closed Principle, created by Bertrand Meyer in 1988 [7, p. 23], states:

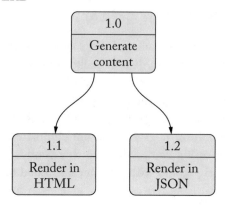

Figure 1.1: Single responsibility principle.

> *A software artifact should be open to extension and closed for modification.*

That is, you should be able to modify the behavior of an artifact (class or other form of abstraction) without having to modify it. There are good reasons for this principle, which arose out of many, many experiences of "simple" modifications to the behavior of a piece of software to fit a new situation breaking the software for all of its previous uses. Modifying software for a local reason is dangerous, and gets more so the more it is used by other parts of the system. Modifying third-party software can be fatal.

1.5.3   LISKOV SUBSTITUTION

In 1987, Barbara Liskov introduced a principle to guide the creation of subtypes (she attributes the actual creation to an unpublished thesis by Gary Leavens), known as the Liskov Substitution Principle [8], which states:

> *What is wanted here is something like the following substation property: If for each object o1 of type S there is an object o2 of type T such that for all programs P defined in terms of T, the behavior of P is unchanged when o1 is substituted for o2, then S is a subtype of T.*

This requires translation, and it helps to come at it from the opposite direction. Basically, the principle says that an instance of any subtype of a given abstraction should be accepted where an instance of the abstraction is required (Figure 1.2). In object-oriented programming, a subclass cannot violate the expectations of its base class, such as not recognizing a message the base class should handle. A similar restriction applies to components that take on other forms.

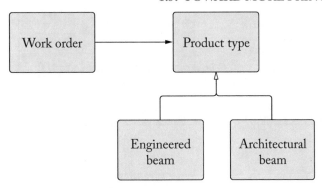

Figure 1.2: Liskov substitution principle.

The principle applies to subtyping, or creating gen-spec relationships that represent an "is-a" concept (an SUV is a kind of motor vehicle). Not all instances of inheritance implement gen-spec relationships, though, so it remains perfectly acceptable to inherit shared functionality.

The reasons for this principle might not be clear. The Work Order in Figure 1.2 accepts an abstraction with a specific interface, called Product Type in this case. The figure shows two subtypes of Product Type, Engineered Beam and Architectural Beam. Following this principle allows you to create a third or fourth subtype that will work with Work Order without modifying Work Order. You have limited the scope of a change.

1.5.4   INTERFACE SEGREGATION

The basic idea behind this principle [1, pp. 119–123] is that the implementation of an abstraction should be separated from the interface to that abstraction (Figure 1.3). This principle originated in the C language in which the interface to a module (the list of methods and their signatures) is published in the header file (.h) while the implementation is published in the source file (.c). In C++ and other languages, we do this by creating an abstract class that defines the interface and is inherited by the concrete class that implements the actual abstraction. Note that this principle applies the Bridge design pattern from [9].

The main reason for this extra effort is to decouple the users of an abstraction from changes to the implementation of that abstraction, leaving them vulnerable only to changes in the interface (which can't be helped and happen far less often). If you as the architect focus on the interfaces between your components and merely describe the components as black boxes, you can limit the degree to which the builders of those components can break other components. This is important, even if you build all the components yourself. Violating this principle can have real-world costs.

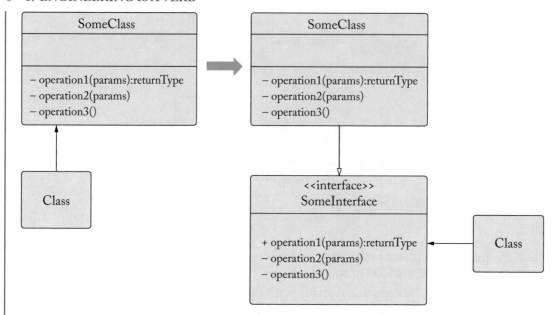

Figure 1.3: Interface segregation principle.

1.5.5   DEPENDENCY INVERSION

Robert Martin [1] says the most flexible designs are those in which source code dependencies refer only to abstractions. Intuitively, an abstract class implementing an interface depends upon the source code that implements the concrete abstraction. The principle says we should invert that dependency, which means a couple of things.

1. Modules that make use of code in another module should refer only to the definition of the abstract interface.

2. Consequently, a module should only import abstract interfaces (*include* in C and C++, *import* in Python, and *use* in Java) of the modules it uses.

It can be hard to live up to this principle. One widely recognized exception is the importing of concrete components from operating system services. When dealing with third-party library code, you have no control over whether you import the abstract interface or the concrete implementation, so it's best to treat these as if they were operating system services (see also the Open-Closed principle in Section 1.5.2).

1.5.6   ONE COPY OF DATA

If you have more than one copy of something in your data, one of them will be wrong. This truism, born out of hard experience, leads to the principle to keep only a single copy of each data element in your data store. The relational data model excels at this, but other types of data stores just take a bit more work.

One of my first full-time jobs was working at Boeing on a replacement for the accounts payable system in one of their major divisions. At the time, a purchase contract was central to four separate systems: accounts payable, the procurement system, the dock receiving system, and the material requirement planning system. The "database" for the accounts payable system was a reel of tape in the data center, and we saw it as a rather large shoebox of microfiche, one sheet per purchase contract. The other three systems ran on the mainframe and used online databases.

Purchase contracts originated in the procurement system and were passed to the other three systems as feeds. So, we had three different databases *on the same mainframe* with a complete copy of the purchase contracts, including history, and one copy on tape. Out of curiosity, we created a report that compared the data for each contract across the databases, listed by contract number. All 4 copies matched only 16% of the time; 3 copies matched 24% of the time, and 2 copies matched only 35% of the time. We had a data consistency nightmare and no easy way to fix it.

This principle was born in that moment. I have since seen it used independently by others many times, so data consistency is a common problem.

1.5.7   KEEP BUSINESS LOGIC IN ITS PLACE

Business logic belongs in the code, specifically in the controllers and models that support the views. Keeping your layers separate will make your life easier in the long run.

When high-end relational database management systems first hit the market, there was a movement to use stored procedures and triggers in the database for business logic. Granted, these tools were easy to use; turns out they were too easy. When we first built the case study application, we made heavy use of stored procedures. We solved the customer credit (many-readers-one-writer) problem (Section 4.12) using triggers. Then things started getting weird.

Putting business logic in the database violates the Layers pattern, but there are also practical arguments against it. Stored procedures and triggers are impossible to manage under configuration control.

The trigger used to solve the customer credit problem worked only when it was on the database assigned to write updates to the customer credit balance, and nowhere else. It kept cropping up where it didn't belong and caused repeated problems (we could never prove that replication wasn't at least partly to blame).

Don't get me wrong, stored procedures are useful tools, but they should be limited to work that belongs in the data layer of the application. I use them to run standard complex queries that we use for a number of things, but they always produce a set of rows or a single value used by some other query. I haven't used a trigger in years.

1.6   WHAT MAKES A DESIGN "GOOD"?

Engineering is a sequence of choices, some easy, most not so much. Design decisions can become much easier, and more reliably made, when you have a defined set of design criteria and some objective way to evaluate them. If more cohesion is better than less cohesion, it would help to be able to tell when a design as more or less of it. I rely on eight primary criteria; many are well known, but I redefined several and others are my own invention. All are described in great detail in *Object Oriented Design Measurement* [5] and summarized in the following subsections.

I have actually been told that measures are "too technical" by people who call themselves "software engineers." Anyone who doesn't want to use measurement because it's too hard cannot be an engineer of any kind. Others see math and freak out. The equations in the following sections are the result of painstaking, detailed work documented in [5]. The equations are easy to use and the text explains what to look for and how to use them. Yes, measurement is technical, but so is software development. A little rigor won't hurt, much, and you'll be better for it.

Abstractions exist in the problem domain and are implemented by design components, which may themselves be single abstractions or a collection of abstractions. Abstractions also arise in the design itself. An array is an abstract type that includes data and expected behaviors, which together we call *properties*. There are many ways to implement an array, but all that matters to the user is that it supports the behaviors expected of an array.

The use of the Model-View-Controller pattern also creates design abstractions, at a minimum one model, one controller, and one or more views. Each of these may be single abstractions or whole frameworks of abstractions.

A key characteristic of any set of problem domain or design abstractions is that together they possess at least those properties of the thing they represent to be sufficient in your design. When you examine a problem domain, you often notice multiple points of view which appear to have different sets of properties for the same thing. For example, the set of properties of a Customer can be very different when viewed from Credit than from Sales or Accounts Receivable. To discover all of the properties of a thing, you must look at it from all points of view. Not all points of view are relevant to an application, or even to an enterprise, so "all points of view" is really "all *relevant* points of view." In most cases, your application cares about only one two points of view.

1.6.1   SUFFICIENCY

Sufficiency can be defined as fitness-for-purpose from a single point of view [5, pp. 368–376]. Does your problem domain abstraction possess sufficient properties to serve the purposes of your application? Do your design components possess sufficient properties to serve as a proxy for the problem domain abstraction in your design?

Compute sufficiency as:

$$suff\,(x,\,a) = |\,x \cap a\,|\,/|a|, \tag{1.1}$$

where x is your design abstraction and a is the reference in your problem domain, $|x \cap a|$ is the cardinality of the set (count of members) of properties in both x and a, and $|a|$ is the cardinality of the set of properties in a. When x is fully sufficient, $|x \cap a| = |a|$ and the measure is one. That is, x is fully sufficient when it contains all of the properties of a. It is possible that x might have properties not in a. This won't affect the sufficiency of x, but may affect its cohesion, as we shall see.

1.6.2   COMPLETENESS

Simply put, completeness measures sufficiency from more than one point of view [5, pp. 377–383]. Of course, it's not *that* simple. A complete abstraction is sufficient from enough points of view to be reused wherever needed in the enterprise. The properties observed from any one point of view usually have a significant overlap with those observed from most other points of view, so the effort to make a sufficient abstraction complete is usually incremental.

Completeness is measured by:

$$cmpl\,(x,\,a) = |\,x \cap a|\,/|a|,\qquad(1.2)$$

where $|x \cap a|$ and $|a|$ are the same as for sufficiency, but are larger because they come from more than one point of view.

1.6.3   COHESION

Cohesion is best expressed as singleness of purpose [5, pp. 384–393]. As Grady Booch put it [10, p. 137], "The class *Dog* is functionally cohesive if its semantics embrace the behavior of a dog, the whole dog, and nothing but the dog." This view of cohesion requires that we define "purpose."

Highly cohesive components are easier to reuse than less cohesive components. An abstraction that is both complete and cohesive can be used anywhere it is needed. In my world of multiple web-based applications that support the same business operation, I have taken to sharing domain abstractions across applications, and over time they have become more and more complete, but have always been cohesive.

Unlike many design criteria, cohesion is pretty easy to assess but not measure empirically. When looking at singleness of purpose, for example, this simple question is often good enough: does this abstraction contain any properties that belong to another abstraction? If the answer is yes, then the abstraction is not cohesive and another look is warranted.

To actually measure cohesion, use:

$$coh\,(x,\,a) = 1 - (|x\backslash a|\,/\,|x|),\qquad(1.3)$$

where $|x\backslash a|$ is the cardinality of the set of properties in x that are *not* in a and $|x|$ is the cardinality of the set of properties in x. Since a value of zero means that none of the properties in x are in a, while a value of one means that x contains only properties that are also in a, you have to subtract the calculation from one to make the measure behave the way we want.

1.6.4 COMPLEXITY

Complexity [5, pp. 330–350] is the most studied, yet least understood, concept in software. Efforts to define complexity fall into two camps: those who define it in terms of the level of effort required to interact with a software component, and those who define it in terms of the component's structure.

More complexity makes everything more difficult. Components take longer to build, they're harder to understand, they interact in strange and unforeseen ways, and if they're complex enough, you just want to ignore them. Since you can only add complexity as you design, you want to carefully monitor how much you add. In practice, this means that when choosing between two sufficient design options, the less complex of the two is usually best.

In my experience, complexity is best viewed as the density of connections between components. It doesn't matter what kind of connection, they all contribute to complexity. Complexity occurs on two levels: the whole design and within individual components. When you measure complexity of design, you care only about the relationships between its components, not the complexity of the components themselves.

Measuring the complexity of a design component or abstraction is much more difficult, but there is a short cut. Because you look only at the connections between components in a design, you can view the design of an abstraction or design component as if it were the entire design. This is different than how [5] defines the complexity of a component, but remember, all you do with complexity is choose among options, and the density of connections within an abstraction or component is good enough.

Thus, for either a whole design or one abstraction (in the problem domain) or design component (in the design), which we designate as x, complexity is computed by:

$$complx_d(x) = |conn(x)|/|x|, \tag{1.4}$$

where $conn(x)$ is the set of the connections between components and $|x|$ is the number of components. This measure falls into the range $[0, \infty]$, although it never actually reaches infinity. Obviously, larger values are worse than smaller.

1.6.5 COUPLING

Coupling [5, pp. 351–367] is like the Force: it permeates your design and binds all components together. Star Wars references aside, coupling is another of those characteristics you cannot eliminate. Coupling is very good at predicting the scope of a change. Dependencies follow lines of coupling: when one object is changed, all objects coupled to it may also need to be changed. Components that are tightly coupled often change together and cannot be reused separately.

Coupling is also easy to measure. While cohesion focuses on the *logical* or *semantic* connections between components in a design, or between the components within an abstraction, coupling focuses exclusively on the *physical* connections between design components. Don't worry

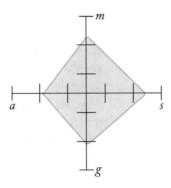

Figure 1.4: Coupling spider diagram.

about coupling within an abstraction—the main reason for creating abstractions is because the state variables (attributes) and operations (methods) are tightly coupled and belong together.

Abstractions in a domain (or design) are connected by four kinds of relationships: generalization-specialization or gen-spec, association, aggregation, or messages. Define the types of coupling for our purposes to be these four types of connections. The *degree* of coupling, from "loose" to "tight," describes the strength of the bond along these lines of connection, but only message connections widely vary in the strength of the bond, as we will see.

Our measure of coupling simply counts the connections of each type between any pair of abstractions, whether in the problem domain or the design. The more connections, the more tightly the two abstractions are bound together. Stated formally, our measure of coupling is:

$$cu\,(x) = (a_g,\ a_s,\ a_a,\ a_m), \qquad (1.5)$$

where each term is the count of the number of connections of each type, regardless of direction. You can draw a picture of the total coupling as a four-dimensional spider diagram (Figure 1.4) and look at the area of the resulting shape.

To measure the relative coupling of two components, you compare the areas of the two spider diagrams. The component with the smaller area has less coupling.

The direction of the connection determines the direction of dependency. In a gen-spec relationship, the generalization should be unaware of its specializations, but every specialization is sensitive to changes in the generalization, so the dependency flows from the specialization to the generalization. This form of coupling is very tight, meaning it is nearly impossible to use a specialization without its generalization. You can mitigate this to some extent by replacing the gen-spec relationship with a form of delegation. Take a look at [9] for patterns you can use.

Aggregations depend upon their components. In an aggregation, one abstraction is effectively an attribute of another. The component aggregation might not know that it participates in such a relationship (it usually does), but there is no way the aggregation cannot know. Changes in the component may require changes to the aggregation, if that means only that the code for the aggregation is recompiled.

In a 1:m association, the dependency flows from the "m" side to the "1" side, the opposite direction of an aggregation. Participants on the "m" side know only the instance of the "1" side to which they are attached, and little else, while the "1" side is usually unaware there is a relationship until you follow the link. Both sides can be isolated from changes in the other, if you implement the association correctly. If you use containment where you make instances of the "m" side attributes of the "1", you have effectively coupled the "1" side to the other more tightly (you've also turned a mere association into an aggregation and flipped the direction of dependency).

Message connections are where things get messy. There are several types of coupling for messages and some types of coupling are "better" (less harmful) than others. You can often trade one type for another.

Because a message from one abstraction to another is literally a function call, message connections vary in the strength of that connection according to the types of module coupling first described by Myers [11]. Myers' forms of coupling form their own ordinal scale, from better to worse.

1. No coupling: Two methods have no connection, in our case, no messages are passed.

2. Data coupling: Two methods communicate via parameters or a set of the same kind values (an array) that contain no control information.

3. Stamp coupling: Two methods accept the same record type or structure, which may contain different kinds of values. This type of coupling creates an interdependence between the two methods that can be very hard to find.

4. Control coupling: Two methods communicate through parameters which have the intent for the sending method to control the operation of the receiving method (control parameters or flags).

5. Common coupling: Two methods communicate through global data. This form is like stamp coupling because both methods depend on the structure of that global data, but they are also coupled by the value of that data.

6. Content coupling: The sending method knows the contents of the receiving method and may branch into or alter the code.

Obviously, you want to limit your message connections to data coupling, this is just good design practice, but sometimes you can relax and use a little control coupling. In Python and other languages, optional parameters are a form of control coupling because the receiving method has to determine whether the parameter was included and react accordingly.

Stamp coupling is valid when you have "packets" of information that always go together and there are multiple types in the packet. Just know that you've coupled both the sender and receiver to the structure of the packet, which may be fine if that's why you created the packet in the first place.

Common coupling is often used to maintain session data and can be the best option in such cases. The alternative is to pass the information in *every* message, which might mean reauthenticating the caller every time. A central database is another form of common coupling, but the data is partitioned so that one abstraction is aware only of its own partition. There is no good reason for any other kind of coupling.

1.6.6   PRIMITIVENESS

A method is either primitive or not; there is no in-between [5, pp. 394–400]. An abstraction or design component is primitive if most of its methods are primitive. Primitiveness and elegance go together.

Primitive methods are shorter, more cohesive, and easier to extract and reuse; they suffer from fewer side effects and unintended consequences [12]. A primitive method usually has lower coupling and complexity, and higher cohesion than its non-primitive counterparts. A method is primitive if it cannot be implemented without direct access to the internals of an abstraction or constructed out of a sequence of other methods of the abstraction.

Primitive abstractions and design components are usually more cohesive and indicate that you have managed to capture the true characteristics of the abstraction. There may be good reasons for an abstraction to not be primitive, but it is usually worth another look. Sometimes, you have no choice: a design component that coordinates the activities of other components, such as the kernel in a microkernel architecture, is not likely to be primitive, but may be as primitive as it can be and still fill its responsibilities.

This suggests that primitiveness is best used when comparing alternatives and not necessarily something to be strived for as an overall characteristic. When you have two or more ways of doing something, the more primitive will most often be the better choice, but you might not have many primitive options.

Berard [12] and Booch [10] define five types of primitive methods.

1. Modifiers: An operation which changes the value of an attribute.

2. Selectors: An operation which accesses but does not change the value of any attribute.

3. Iterators: An operation that permits all parts of an abstraction to be accessed in some well-defined order.

4. Constructors: An operation that creates and initializes the state of an abstraction.

5. Destructors: An operation which destroys the abstraction and frees the resources it consumes.

A method that uses one or more selectors is primitive if no messages are sent to other components. Sometimes, multiple attributes have to change together to be a valid state change, so a method that causes a single valid state change using modifiers, with no side effects (messages to other components), is primitive. A method that uses an iterator, constructor, or destructor cannot be primitive.

For the abstraction, since a method is either primitive or not, the primitiveness of an abstraction can vary from not at all (no methods are primitive) to fully primitive. This is easy to map to the range [0, 1], and our measure for the primitiveness of an abstraction is:

$$prim(x) = \frac{\text{number of primitive methods}}{\text{number of methods}}. \tag{1.6}$$

1.6.7   SIMILARITY

You don't see much written about similarity [5, pp. 401–413], but we use it all the time in design. Object-oriented designers will often combine two separate classes into a single gen-spec structure by factoring out the common function or structure as a way to reuse code. My use of mix-ins (Section 4.3.4) to implement role-based access control in the case study is an example.

There are four types of similarity: structural (with four subtypes), functional (with two subtypes), behavioral, and semantics (purpose or intent). Including subtypes, there are eight cases, each of which provide clues as to how you might reduce the code you write by changing your design.

Two abstractions are *structurally similar* when they have similar structures, which isn't as obvious as it sounds (structures can look different and still be similar, or look similar but not be). Similarity of structure can be observed *internally* or *externally*. External similarity, the first subtype, can be observed directly, but internal similarity can be hidden by the names of attributes and operations. Further, the same operation may be implemented differently in two abstractions, so you can't rely on the sets of methods, either.

To judge internal structural similarity, you have to boil the apparent structure down to its essence to determine if you have two different components, or the same component implemented differently. If an attribute of one component is the same type as an attribute in another, they are similar, even if they have different names. This is the second subtype of structural similarity, call it *internal attribute structure*. If all of the attributes in a component are the same types as all of the respective attributes in the other component, the two components are similar in the structure of their state.

The third subtype of structural similarity is *internal method structure*. Two methods are said to be structurally similar if their signatures have the same number and types of parameters, even if the order is different. Two components are similar if all of their operations are similar.

Two components can participate in the same set of relationships with the same set of corresponding components, for the fourth subtype of structural similarity, called *relationship structure*. In this case, the two components may well be two subtypes of the same thing, but you have to apply Liskov's substitution principle to be sure. If they are, you can save some code by creating a generalization of the two and moving the relationships to the generalization. This might not help now, but it is still a good decision for the future.

Structural similarity is very weak and is not often a good reason to combine two abstraction, even if you combine them into a gen-spec structure. Even though the physical structures might be similar in terms of types and counts, they may not be at all similar in meaning and combining them could very likely reduce cohesion and increase coupling, both bad outcomes.

Two methods are *functionally similar* if they have the same effects (*physical function*) or apply the same logic (*logical function*). The effects are determined by the differences between the pre- and post-conditions of the method; they do the same thing but do it differently. The logic is internal to the method itself; they do the same thing but in different contexts.

Methods that have the same effect can often be grouped into structures using a strategy or decorator pattern [9]. Methods that have the same logic can often be reused by factoring them into a stand-alone abstraction and moving it to the application-generic domain.

Two components are *behaviorally similar* if they have similar lifecycles or respond to similar events in a similar way (including imposing similar sets of pre- and post-conditions). Similar lifecycles are most often the result of similar structures, so look there first. Behaviorally similar components can often stand in for one another, and it is rare to actually need more than one unless there are compelling reasons for it. Never pass up a chance to write less code if you can get away with it.

Two components are *semantically similar* if they exist for the same reason, and one can be used in place of the other. Two components that are sufficient but not complete and represent different views of the same domain concept are not similar; one cannot be used for the other, but they can be combined to create a new abstraction that is still sufficient but more complete. Semantic similarity is a subjective judgment, much like the choice of using one word or one of its synonyms in a sentence.

The measure for similarity is a seven-dimensional measure constructed one dimension at a time using each of the seven types of similarity (all of the above except semantic similarity). They all take the same general form, and are all comparisons of sets of properties. Further, since they are comparisons, all you can tell is how similar two abstractions are; you cannot apply this to a single abstraction (which doesn't make any sense, anyway). Formally, each dimension is defined as:

$$sim\,(a,\,b) = |a \cap b| - |a \backslash b| - |b \backslash a|, \qquad (1.7)$$

where a and b are, respectively, the sets of attributes, operations, external structures, physical function, logical function, events, and event responses for two abstractions (or two methods in the case of physical and logical function), $|a \backslash b|$ is the number of properties in a that are not in b, and $|b \backslash a|$ is the number of properties in b but not in a.

1.6.8   VOLATILITY

Simply put, volatility [5, pp. 414–421] is the likelihood that a design component will have to change. Volatility comes from two sources: the problem domain and changes to other components.

Volatility in the problem domain also has two sources: changes to the problem domain itself, and changes to the enterprise's understanding of that domain. One is part of the environment and must simply be dealt with, the other can be dealt with, but not by you. From your perspective as a designer, they are the same.

Volatility from other components follows lines of coupling and can create volatility where none actually exists. Dependence is one type of coupling and is a necessary part of the design. A specialization is dependent on its generalization, an aggregate is dependent on its components, the sender of a message is dependent on the signature of that message in the receiver's public interface.

Exposure is another type of coupling and is the result of sloppy design. You can almost always insulate a component from changes in other components, even those upon which it depends, although it isn't always worth the effort. In the case of messages, there are techniques to decouple the sender and receiver, with the tradeoff that the dependency is transferred to some other component.

Exposure is the probability that a component will have to be changed because of a change to some property in another component. Exposure follows lines of coupling, and the degree of exposure depends on the type of coupling, but a coupling relationship does not necessarily increase exposure.

Exposure can be physical or logical. Logical exposure lives in the design and is the result of changes to the design of a component, either in terms of structure or behavior. The results of a change can range from having to modify the code to a simple recompile to no effect at all. One of your design objectives is to try to limit the effects of changes to the least painful. Physical exposure has many of the same characteristics and effects of logical exposure. We deal with physical exposure in Chapter 5.

Volatility from the problem domain is something you have to guess at. The more experience you have, especially in the problem domain and context in which you're working, the better feel you will have for where volatility exists. In data-strong applications, volatility lives mostly in the business logic and user interface. In function-strong applications, the mathematics applied in the filters will change most often. In control-strong applications, volatility comes from everywhere; one of the main purposes of the microkernel architecture is to isolate one part of the system from all other parts.

You can measure the volatility from the problem domain as a straight probability value in the range [0, 1] and denote it as likelihood of change, or $loc(x)$.

Exposure is also a probability, but combines the likelihood a coupled component will change and the kind of change that is likely. Because you must at least examine your component when a coupled component changes, you are really interested in the volatility of that coupled component. Because your component can also be coupled to multiple components, the complete exposure is the sum of exposures to each component, which in turn is the sum of the volatilities $vlt(k)$ of the properties of a component that is coupled to yours. Given a component x, the measure of exposure $exp(x)$ is:

$$exp(x) = \sum_{k \in A} \sum_{i=1}^{n} vlt(k_i),\tag{1.8}$$

where A is the set of components in the design.

Combining the two, we define $vlt(x)$ as:

$$vlt(x) = loc(x) + exp(x).\tag{1.9}$$

The combinations of Equations (1.8) and (1.9) says that volatility can be experienced recursively. If component A is coupled to components B and C, which are coupled to components D and E, and F and G, respectively, a change to component G can affect component A, depending on the effect the change has on component C. Volatility is a creeping mold for which you need to be on constant watch.

1.7   A WORD ABOUT OUR CASE STUDY

Throughout this book, we will use a variation of a project for which I was the lead engineer in the late 1990s. The application ran the front office of a wholesale building materials distribution company that turned between \$15B and \$16B in annual revenues. The application handled sales, purchasing, inventory, custom work orders, and load planning and logistics. It interfaced with the corporate finance system at several points, including a rather complex mechanism for figuring out the account numbers to use for various business transactions (this could just as easily have been part of the financial system, and probably should have been, but we were asked to implement it, so we did).

I will present the project as I would build it today, as a web-based application using Django (and Python) or Rails (and Ruby). My last four projects have all been web-based applications, the last two being built in Django and Python. Three of these projects run on MySQL databases, the fourth on SQLite because it was my only option that didn't require the use of a full-time database administrator (that was an imposition by the company). All four involved HTML, JavaScript, and SQL.

Today, most web-based applications follow the same general architecture. They are generally data-strong applications (Section 2.3.1) that follow the basic shape of the Blackboard pattern and use the Application Controller [13] and Layers [14] patterns in the implementation.

Like most web frameworks, Django follows the Model-View-Controller (MVC) [14] pattern to a point. Django has models, views, and templates, with Django views equating to

MVC controllers and Django templates equating to MVC views. The addition of JavaScript on the front end means that the MVC pattern is replaced with the Application Controller pattern [13].

As we will discuss in Section 2.4.1, data-strong applications lend themselves to object-oriented programming. This is by no means a requirement, but life is much, much easier since the class model largely follows the data model, with additional classes added during design. In Section 2.4.2, we will see why I used the data-driven design approach for most of the application.

1.8 SUMMARY

Engineering is a verb. While that's more repetition than summary, it's worth repeating. You *do* engineering, regardless of what you *call* yourself. Conversely, if you don't do engineering, it doesn't matter what you call yourself.

Design and engineering have a complicated relationship. Engineering requires design, providing a discipline and rigor that has been lacking in software development. As part of the design process, engineering requires and provides means to analyze and evaluate designs. Engineering requires the use of measurement. In this chapter, we explored a number of principles to provide discipline, and design criteria, with ways to measure them, to use for analysis and evaluation.

Applying these to your designs will make you a better designer, and lead you to create better software, even if you don't write software for a living.

1.9 FURTHER READING

Some of my favorite sources on the nature of engineering are from Henry Petroski, Professor Emeritus at Duke University. Petroski's main theme is on the role of failure in design and engineering. In addition to [3], here are some books that provide often entertaining insight into the history and development of engineering as a discipline.

- Petroski, H. (2006). *Success Through Failure: The Paradox of Design.* Princeton, NJ, Princeton University Press.

- Petroski, H. (1996). *Invention by Design: How Engineers Get from Thought to Thing.* Cambridge, MA, Harvard University Press.

- Petroski, H. (1994). *Design Paradigms: Case Histories of Error and Judgement in Engineering.* Camebridge, UK, Cambridge University Press.

- Petroski, H. (2003). *Small Things Considered: Why There is no Perfect Design.* New York, Vintage Books.

This text on the mathematics of software engineering and the following text on object-orientation in Z (pronounced "zed," a formal mathematical notation for programming) provided the foundation and impetus to develop a calculus for analyzing software designs, presented in Chapter 6 of [5].

- J. Woodcock and M. Loomes, *Software Engineering Mathematics*, Addison-Wesley, Reading, MA, 1988.

- S. Stepney, R. Barden, and D. Cooper, *Object Orientation in Z*, Springer-Verlag, London, 1992. DOI: 10.1007/978-1-4471-3552-4.

Here are some general software engineering sources I have found useful over the years.

- Boehm, B. W. (1981). *Software Engineering Economics*. Englewood Cliffs, NJ, Prentice Hall.

- Leach, R. J. (2000). *An Introduction to Software Engineering*. New York, CRC Press.

- Humphries, W. S. (1995). *A Discipline for Software Engineering*. Reading, MA, Addison-Wesley.

- Glass, R. (1991). *Software Conflict*. Englewood Cliffs, NJ, Yourdon Press.

- Pressman, R. S. (2005). *Software Engineering: A Practitioner's Approach*. New York, McGraw Hill Higher Education.

- Brooks, F. (1975). *The Mythical Man-Month*. Reading, MA, Addison-Wesley.

Here are some good resources on software measurement, in addition to those in [5].

- Hubbard, D. W. (2010). *How to Measure Anything*. Hoboken, NJ, John Wiley & Sons.

- Fenton, N. E. and Pleeger, S. L. (1997). *Software Metrics: A Rigorous and Practical Approach*. New York, International Thomson Computer Press.

- Basili, V. R., Caldiera, R., and Rombach, H. D. (1994). Goal question metric paradigm. In J. E. Marciniak, *Encyclopedia of Software Engineering*, vol. 1. New York, John Wiley & Sons.

Finally, Agile, as practiced today, is more about organizing and managing software projects than about actually building software. Still, some of the earlier sources covered actual development. Here is one that I liked, from one of the originators of Agile development:

- Beck, K. (2000). *Extreme Programming Explained: Embrace Change*. New York, Addison-Wesley.

CHAPTER 2

Architecture Matters

All architecture is design; not all design is architecture. The difference? If you have to start over to fix it, it's architecture.

Remember those story problems in high school algebra? A lot of people hated them; I loved them. The hardest part was interpreting the story to get the equation. Get the right equation, the algebra was (mostly) easy. Get the equation wrong, you spent time and effort *solving the wrong problem*, with no hope of getting the right answer. Architecture is like that: get it wrong, you have to start over. The most important skill in architecture, of any kind, is recognizing the story; sometimes, you have to write it. The purpose of this chapter and the next is to help you make sure you solve the right problem. You may still get the wrong answer, stuff happens, but at least you were on the right track.

In the mid-1980s, we had two options for hosting an enterprise application: minicomputers and mainframes. Personal computers were just coming into their own and would pretty much change everything over the next few years, but in 1986, mainframes and minicomputers were our options. The team I was on had a lot of discussions about how to decide between them. Mainframes were reliable, highly available (were up most of the time), could handle large numbers of users over a wide geographic area, but were very expensive to use. They were also very hard to develop on. Minicomputers were far cheaper, but tended to be slower and crashed more often. Their lack of speed meant they couldn't support large concurrent user communities. We finally settled on a set of questions to determine which platform to use:

1. How many users are there?

2. How many time zones do they cover?

3. How is the data generated and used?

4. What are the availability requirements?

The selection of the hosting platform went a long ways toward determining the architecture of the application. Mainframe applications are structured differently than those hosted on minicomputers, for a variety of reasons. The high cost of mainframe applications limited the extent to which the user interface could be online, and even when those applications were online, the user interface was less than ideal. Think "green screen" (where do you think the term came from?). Minicomputers offered much more flexibility in how applications were designed and built, but the platform suffered from limitations.

Today, we make decisions about whether to host the application in-house on one of our own servers, or in the cloud. We make separate decisions about where to put the user interface—on the web, on a custom-built front-end application, or on a mobile device, or even all of the above. These decisions drive the architecture of the application. In fact, these decisions *set* the architecture of the application, well before we begin to decide what the application needs to do.

As important as architecture is to the success of an application, this is not a book about architecture. Most books about architecture actually talk about how to *document* an architecture, or how to *organize the practice* of architecture. We're not interesting in either of those; we want to know how to *identify* an architecture so we can build a well-engineered system that can successfully solve the problem at hand.

Note that I use the word *identify* rather than *design*. The truth is, most of the architecture of any system, or any structure for that matter, is determined by the nature of the problem and the context in which the solution will exist. This may sound a counterintuitive, but you'll understand by the end of this chapter.

2.1   IT'S ALL ABOUT THE PROBLEM YOU ARE SOLVING

All software is written to solve a problem, to serve some purpose. Like any other tool, software is judged by a single criterion: fitness for purpose. That is, can it be used effectively for the intended purpose? If not, the tool is an abject failure. In the long run, only fitness for purpose matters; no one will remember how much it cost or how long it took to deliver.

That purpose or problem, then, is the primary source of information for what our tool needs to be and do, and that, in turn is the primary driver of the form our tool will eventually take. The problem alone is insufficient, however, as the context in which our tool will be used also has a significant influence over its form.

Architecture is a design activity. An architecture describes the basic shape of a solution, including the purpose for which it will be built. The phrase "form follows function" is among the first lessons taught to architects and indicates the primacy of fitness for purpose. An architecture is thus a designed solution. All architecture is design, but we will see that not all design is architecture (the original version of this quote is attributed to Grady Booch in a 2006 article that is no longer available).

Architecture requires skill in both analysis and design. The difference between them is the desired outcome. In analysis, we use deductive reasoning to take something apart, a problem in our case, so we can better understand it. Design, on the other hand, uses inductive reasoning to synthesize or build something that didn't previously exist.

In this chapter, we do a lot of both. We analyze parts of the problem to understand them in order to build better models that will feed the rest of the design process. This is by far the most critical activity for the eventual success of our effort. Getting this wrong generally requires starting over.

2.2   CREATE A PATTERN LANGUAGE FOR YOUR SOLUTION

Christopher Alexander, in his ground-breaking work on patterns in architecture [17], talks about creating a *pattern language* for a project, a list of high- and low-level design patterns that work together to create a whole architecture for the solution. The creation of this pattern language is the single most important task for the architect, but is surprisingly simple. The practice of designing the architecture of a solution is largely one of finding patterns to solve various parts of our problem, and using them to craft the solution.

Starting with the shape of the problem, work through the non-functional requirements discussed in the sections that follow to build your pattern language.

2.3   WHAT INFLUENCES ARCHITECTURE?

When designing a bridge, the biggest influences over the type of bridge—its architecture—are the nature of the crossing and the type of traffic it will carry. The length of the span is the most influential, as longer spans constrain the types of bridges you can build. Up to about 100 feet, any type of bridge will work, and cost becomes the biggest design constraint. Between 150 and 300 feet, some types of bridges no longer work, limiting your options. Between 500 and 1000 feet, there are basically four options, and the nature of the crossing itself becomes the limiting factor. Beyond 1000 feet, you are down to two or three options that differ by where the weight of the bridge will be supported, which is influenced by the nature of the bottom and ends of the crossing. In these long spans, if the crossing is such that piers can't reach the bottom, no bridge may be feasible.

Likewise, the predominate shape of the problem has the most influence over the architecture of a software solution. Once you identify that shape, you've gone a long ways toward defining your architecture. But you're not done. What else influences the architecture? What other patterns might be needed?

As we saw at the top of this chapter, the context or environment in which the software will run significantly influences the end result. In fact, these *non-functional* requirements have more influence than *functional* requirements. For those unfamiliar with the distinction, a *functional requirement* is a statement of *what* a solution needs to do. A *non-functional requirement* is a statement about *outside* or *environmental influences* on the solution and its operation.

Think about your problem and your intended solution, and look back at the four questions at the top of this chapter:

1. How many users are there?

2. How many time zones do they cover?

3. How is the data generated and used?

4. What are the availability requirements?

Based on my experience since then, I've add a few more questions:

1. What is the basic shape of the problem?

2. Where should things go?

3. What level of scale do you need?

4. How much will that scale change?

5. What are the performance requirements?

6. Where does volatility live in the problem? What is likely to change the most and most often?

7. How much variability do I need to worry about?

8. What about security? How do we handle authorizations and prevent unauthorized use? What are the risks and the consequences of a security breach?

The answers to these questions provide the bulk of your non-functional requirements. We'll talk about each in the following sections, and I will name a few design patterns that I have found useful over the years.

2.3.1   PROBLEMS HAVE A SHAPE

Software design problems have a shape that will significantly constrain your design options. Problems can be placed into one of three classes based on the relative dominance of the three dimensions of software [18]:

- Data: A partitioned view of what the system remembers

- Function: A partitioned view of what the system does

- Behavior: A partitioned view of the different behavioral states that characterize the system

DeMarco claimed that these perspectives, and only these three, are required to describe any piece of software. DeMarco's analogy was the three dimensions required to describe a physical structure. While one could argue the applicability of the analogy, the idea has merit. DeMarco's perspectives are so basic I have taken to calling them dimensions, largely because of his use of this particular analogy. Over many years of discussion and use of these dimensions, I have taken to calling the third *Control* for reasons that should become clear soon.

It turns out that these three dimensions are actually characteristics of the problem domain the software is built to solve. They thus form the top level of a set of patterns we can use to describe the shape of a problem domain.

Some problems are clearly dominated by one of these dimensions and can be classed into one of three categories: data-strong, function-strong, or control-strong [19]. Other problems are considered hybrid because the dominant dimension isn't as readily apparent.

Let's be clear, here. We classify applications based on the *dominant* dimension present in the problem. Every problem, and thus every software application, no matter how large or small, has some aspect of all three dimensions. You will find yourself forgetting this from time to time, usually at your peril.

Historically, software people have grouped applications into three classes: Business, Scientific, and Real-Time or Embedded (real-time is more correct because embedded just describes where the software lives, not what it does). Perhaps not surprisingly, these three classes of software applications align very closely with which of DeMarco's dimensions dominate the problem.

Thus, we classify applications using the shape of the problem. The shape of the application naturally follows that of the problem. This is the first point where a pattern mismatch can occur, but only the first. While you can use a solution of one shape for a problem of another, the results are at best more work, and at worst complete failure.

Data-Strong

Data-strong or business applications model information about concerns that exist in the problem domain: people, concepts, and things. A customer relationship management application (CRM), for example models information about customers, tracks communications with those customers, and helps manage the overall relationship. A *customer* is an object that exists within such a system and may have a very complex structure. To make a CRM work, other concerns are needed, but like customer, the application contains models of information about them.

The data dimension significantly dominates the other two and is the primary source of complexity, along with the operational context for the application [20]. Further, the data model serves as the starting point to design and construct such a system.

Data entities exist in the problem domain as nouns in the business vocabulary. Listen to what people talk about as they conduct business. Odds are, you'll hear words like *customer, employee, location, order, item,* or *supplier*. These are all likely entities in the data model of many applications in that business. Other businesses have a slightly different set of nouns. Some businesses refer to their customers as *clients*, or their employees as *associates*. From a modeling perspective, a customer and a client are the same thing, likewise for an employee and an associate, but the applications for each business may reflect the terms they actually use.

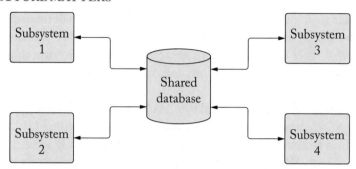

Figure 2.1: Blackboard pattern.

The model isn't just about entities, however, it's also about the relationships between them. Some relationships are inherently structural while others are governed by business rules. A customer, for example, may have one or more addresses that serve various purposes, each important to some part of the company. In many businesses, the very idea of *customer* can lead to a rather complex data model when you consider subsidiaries, who pays for orders, and who the contact people are. The customer-address relationship is structural and represents the notion that a customer *has* or *contains* one or more addresses. An example where business rules determine the relationship is that an employee has only one manager.

The case study application introduced in Chapter 1 is a data-strong application. As in most data-strong applications, the function and control dimensions are non-trivial and require some attention. We will get into this more in the next chapter.

The Blackboard pattern in [14] best matches control strong problems. The centrality of the shared data store is the defining features of both data-strong problems and the Blackboard pattern (Figure 2.1).

Function-Strong

Function-strong or scientific applications model the way things behave in the real world. Put another way, a scientific application models the behavior of a real-world object rather than its attributes or state, although some state information is almost always required. The mathematics or logic that describe the behavior are the source of complexity in such applications.

The function dimension nearly completely dominates the other two. The most common architecture for such applications clearly reflects the mathematical or logical operations that are found in the problem domain.

There are function-strong applications that are not based on behavior modeling. Image analysis, for example, consists of a string of operations performed on an image in order to extract information from it. A typical image analysis pipeline converts the raw image into a two- or three-dimensional array of numbers that represent the level of intensity, and perhaps color, for each pixel. In a 3D image the pixels are called voxels and have a known size in all three dimen-

Figure 2.2: Pipes and filters pattern.

sions. This conversion to a numeric array is the first step in any pipeline and depending on the type of image, can be very complex. Subsequent steps accept the numeric array, perform some set of operations, then output a modified array. Usually, each step performs only one operation, or a short sequence that are *always* done together.

Applying a machine learning model is another example of a non-behavioral filter. Applying a model is usually one step in a multi-step pipeline, nearly all of which follow the same architectural pattern and are clearly function-strong.

Function-strong problems, with their strings of processes, most closely resemble the Pipes and Filters pattern [14] (Figure 2.2). You may recognize the shape in the Named Pipes of Linux shell commands.

Control-Strong

Control-strong or real-time applications are a different kind of animal. Unlike business and scientific applications which model some set of properties of objects in the real world, real-time applications must deal with the world itself on the world's terms. The number and kinds of events that the application must detect and handle and the timing and concurrency constraints contribute the majority of complexity in these applications. Real-time applications are often called embedded because they reside inside other types of products, such as airplanes, cars, refrigerators, and ovens. Not all real-time applications are embedded; however, factory or powerplant control systems being two examples.

The control dimension dominates the other two. In some cases, the data and function dimension are so small as to be hard to find, but they exist.

Control-strong problems most closely match the Microkernel pattern [14] in which a relatively simple kernel manages and responds to events that are detected and handled by separate components. Adding a new event is as easy as adding a new component (Figure 2.3).

2.3.2   WHERE SHOULD THINGS GO?

The early days of mobile apps (think prior to 3G) were a lot like the early days of the Internet: computers didn't have a lot of power, memory was always in short supply, screens were small, and the network was slow and unreliable. It was common to have great cell service until you walked into a big box store.

App designers had to worry about how things would work when there was no network. A deliberate decision had to be made about what had to work no matter what, and provision for that functionality had to be built on the phone. It was quite the dilemma.

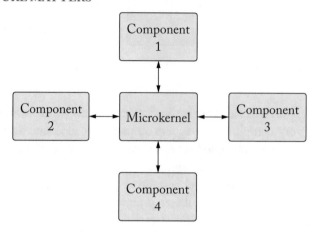

Figure 2.3: Microkernel pattern.

With LTE and now 5G, wireless networks are much more reliable, but the problem hasn't gone away. You also face it when designing web-based applications.

Given some functionality, should it happen on the front end, the back end, or some combination of the two? The front end and back end of both mobile apps and web-based apps are separated by a chasm called "the network," typically with an asynchronous connection. There are tools, such as Web Sockets, that create synchronous connections over the network, but doing so doesn't help that much; a chasm with a synchronous crossing just provides more opportunities to lock the front end waiting for the back end to respond. An asynchronous connection at least allows the user to go do something else. Nygard [6] describes synchronous calls across a network an antipattern, something you should not do.

There are two main differences between the two types of connections, one of which I've already alluded to. In a synchronous connection, the sender waits for the response before doing anything. Program logic is more simple, at the cost of the occasional block. In an asynchronous connection, the sender sends the message along with a "callback" function to execute when the receiver finally gets around to responding. Program flow is much more complicated, but the user is never blocked waiting for the other end (unless you code it that way). If you ever coded an AJAX call in JavaScript, you've seen this structure.

The other main difference is that since the sender doesn't wait for the receiver, there is no guarantee that the state of either side will remain consistent for the response. Asynchronous connections are sometimes referred to as stateless for this reason.

In such an environment, the designer needs to be clear where functionality will occur. It's very easy to get confused when writing code only to discover that the data you need is on the other end.

For a recent application, I built a tool into a web-based application that would allow a user to measure features on an MRI scan. The MRI was loaded on the server side, converted

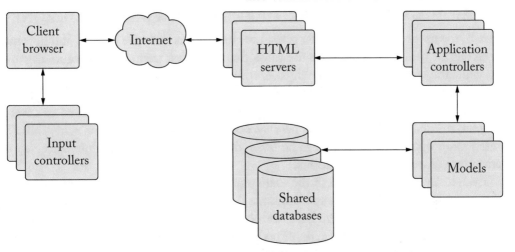

Figure 2.4: Typical web application architecture, a.k.a. Application Controller and Layers patterns.

to a numeric array, then sent to the client. The client had the entire array and all of the logic necessary to display and annotate the image for the feature in question (a tumor, in this case). The results were captured in a sparse array that was sent back to the servers which performed several calculations. This was designed deliberately to minimize the network traffic during the process at the expense of front-loading the client's web browser. For most images, this works just fine, but some images weigh in at just over 5 GB and browsers cannot handle that kind of memory load. The fix will require reversing the trade off and trade space for time due to network traffic. It will also require the back and front ends to more closely coordinate the state of the client (keeping track of which slice is being analyzed, for example).

As applications are asked to do more, designers will encounter these decisions more often. As you can see, throwing more and faster hardware at the problem is less and less a viable solution.

The Layers pattern [14] is the most often used pattern for packaging functionality. Once packaged into layers, they can be allocated to either side of the network, and moved with relative ease. Chapter 5 covers these packaging decisions in detail.

The Model-View-Controller and its variant the Application Controller [13] also provide guidance for packaging functionality. The Application Controller is particularly useful when a network is present and there are control or workflow issues on both the client and server sides of the network. The use of JavaScript in a web application is an example of client-side application control. These patterns serve to lump together those parts of your application that tend to change together most often, isolating them from other parts. The use of both Layers and Application Controller are shown in Figure 2.4.

2.3.3   HOW MANY USERS ARE THERE AND HOW MANY TIME ZONES?

The size of the user community has two parts: how many users there are in total, and how many of those will be using the solution at the same time. The first part creates data management issues, while the second can affect literally everything.

It used to be that a system had to operate reliably during the normal business day, or first shift. A system that provides functionality to just the United States still needed to cover six time zones, so "first shift" was 13 hours per day. For a global website, it's always first shift. This mattered more when the system had to be taken down occasionally for maintenance, another big advantage of web-based applications.

Even today, though, the size of the user community, especially the concurrent user community (users accessing the solution at the same time) matters a great deal to the architecture of the system. In recent years, there have been several spectacular failures when new websites launched only to crash under the weight of people wanting to use them. There are several well-known ways to mitigate these issues which are becoming standard practice, but not fast enough, apparently.

The best way to deal with large or even large-ish user communities is to scale your front end horizontally using load balancers and lots of servers. Load balancers direct traffic where it is lightest and get out of the way. I don't often explicitly use load balancers as most of my applications aren't big enough to need them. Instead, I direct all traffic to the one server that then spreads it around over several server processes running in the background. This all happens within a single container and works surprisingly well.

Had the case study application been built as a web-based application, we would have used load balancers as the user community numbered in the thousands.

2.3.4   WHERE IS THE DATA GENERATED AND USED?

In many distributed systems, such as wholesale distribution, information about things like available inventory is generated and consumed locally. Depending on the cost of shipping items, distribution centers do not often dip into the inventory of another center.

A website selling to the whole world, such as Amazon's Marketplace, has a different problem: they can get an order from anywhere and need to fulfill it from the most cost-effective source, *after* they have confirmed they can fulfill it at all. As with large user communities, there are proven ways to solve this type of problem. All we note here is that it comes in many forms and has little to do with what the solution is actually doing.

Data that is shared across a distributed application can be a particularly nasty problem. There are several published patterns that look like they might apply in such cases, but I haven't used any (yet), mainly because I had solved the problem before the patterns were published. I present one solution in Chapter 4.

2.3.5   WHAT ARE THE AVAILABILITY REQUIREMENTS?

We define availability as whether the solution responds when someone tries to access it. Put another way, what is the cost, to both you and the user, if the system is not available? The answer depends on a lot of things and is unique to every situation, but also has a major impact on the architecture.

Michael Nygard [6, pp. 7–134] describes this as *stability* and provides several design patterns that can enhance the stability of your application. While Nygard calls them design patterns, they have to be baked into your design from the beginning, making them architectural patterns; you can't just bolt them on after final testing shows you have a problem.

Timeouts [6, pp. 100–103] build in a way to disconnect and move on when a remote resource is too slow to respond. While this pattern doesn't solve the problem, it does give the user the option to go onto something else or try again, thereby improving the perceived availability.

Handshaking [6, pp. 123–124] is a pattern that lets a server control the pace of its work by throttling its response time. Based on the idea that a slower response is better than simply crashing, the use of handshaking can also improve perceived availability.

Decoupling Middleware [6, pp. 130–132] describes the use of software that sits between the sender and receiver of messages, ensuring the receipt of the messages in the order in which they were sent, while allowing the sender to go on their merry way. Middleware tools can range from very simple to very complex, and include collection of tools known as the Enterprise Service Bus.

Information technology folks tend to measure availability in terms of the percentage of time a system is available, usually called "the nines" referring to the number of nines in the percentage, with 90% being one nine, and 99.99% being four nines. Historically, each additional nine results in a 10-fold increase in cost, especially when measured at the data center.

The first 4 nines, to 99.99% are fairly easy, as an availability at that level means about 52.6 minutes of downtime *per year* (3 nines, or 99.9% equates to about 8.77 hours per year). The first 4 nines can generally be achieved by using better hardware, or more hardware. The virtual servers running the case study application have not crashed in over five years of operation.

The gold standard for availability is 5 nines, 99.999%, or 5.62 minutes of downtime per year. That level of availability affects the architecture in dramatic ways. Not only do you need redundant hardware, with multiple servers running *every* component of the software, but those servers needs to be located in multiple physical locations, preferably in multiple climates (see Section 5.3.2 on Redundant Geography).

Early in my time at T-Mobile, they had just completed a new data center after having the first one flood out for three days (think about the cost of a mobile phone carrier being down for three days—in comparison, the cost of a new data center was spare change). The new data center was located above a major river in Central Washington State and directly connected to two of the hydroelectric dams for power. They thought they were redundant, but if the up-river dam ever collapsed, the next dam down river was going to go with it. Redundancy, or the lack of it, can bite you in some very unexpected ways.

2.3.6   WHAT SCALE AND SCALABILITY DO YOU NEED?

Scale is the volume of activity at its *peak* level. If you're a toy store, your peak sales are probably in the fall, and can be many times more than your average sales throughout the year. If you're a power generator, peak demand varies by time of day and time of year, and whether you're in the desert or a colder climate. I currently live in Arizona, where peak demand for power is late afternoons from late-Spring to early-Fall. Scale is the maximum level of expected activity and the solution needs to be able to deal with it. As scale increases, the architecture becomes more complex and expensive. It costs a lot less to deal with 1,000 transactions per day than it does 1,000,000 transactions per day; or 1,000 concurrent users vs. 1,000,000.

Scale is generally measured in orders of magnitude, or powers of ten, because the first number doesn't really matter. The difference between 1,000 and 5,000 is largely immaterial, but 10,000 is a different animal. Design for peak scale.

Scalability is the degree and frequency with which the volume of activity changes over time. Activity for the toy store might be relatively flat for most of the year with a big spike in the fall. The activity for a power company has daily peaks that rise and fall throughout the year.

Nygard [6, pp. 135–208] lumps these two concepts, and the performance described in the next section, into a single idea of *capacity*. Nygard defines capacity as: "…the maximum throughput a system can sustain, for a given workload, while maintaining an acceptable response time for each transaction…." Throughput and response time are both measured from the view of the end user, so capacity is also an end user view of the system. He provides a number of design patterns that can help create capacity without additional infrastructure. Unlike his patterns to enhance stability, his capacity patterns *can* be added after the rest of the system is done, and so are truly design patterns. I've used all of them at one time or another, usually after the system had gone live, and can attest they work. We cover these patterns in Chapter 5.

The way in which activity changes matters to the architecture. The toy store can get by with very little staff most of the time, then hire additional people when needed. Both the store and potential employees know this pattern and can adapt to it. The power company has to maintain peak capacity pretty much all the time. One of your jobs as an architect is to determine the pattern in activity levels and design a solution that is both effective (adapts to the level of scalability) and cost-effective.

2.3.7   WHAT LEVEL OF PERFORMANCE IS REQUIRED?

A long time ago, I read an article about measuring response time that made a lot of sense. At the time, a large majority of requirements specifications included a statement about "sub-second" response time. That is, the system was required to respond to an input in less than one second. That was bunk, to quote the article, as only four levels of response time actually mattered.

- *Eye-blink*: The system responds immediately, which is up to about 2 s, to an action from the user.

- *Delay*: In which the system doesn't respond immediately but responds faster than you can do something else and come back.

- *Wait*: You know it's going to take a while, so you can plan to do something else while the system is responding.

- *Overnight*: We don't see much of this anymore, but one example is waiting for a deposit to clear your account and be made available. We just know it's going to take time, and we build that into our expectations.

I write code for a living, and remember when a compile of a reasonably sized application went from several minutes, wait time, to less than five, delay. I used to start a compile, go get a cup of coffee, maybe talk to the boss, and get back just as the compile was finishing. As computers got faster, so did compile times. It was very frustrating when all you could do was bide your time waiting for the compile to finish because there wasn't enough time to do anything else.

Today's world of rapid-response websites, along with network delays and lags, creates a set of conflicting expectations for users. We've come to know instinctively that if a transaction requires traversing a network, it's just going to take longer. Designers of modern applications have to make conscious decisions about what needs to happen quickly and what can wait. We use visual cues, such as buttons that change color, to tell the user: "We heard you, now give us a minute to get back to you."

The architecture and design patterns that help improve the perceived availability of an application, if not the actual availability, also help with the perceived level of performance. From the user's perspective, *any* action is better than no action, so even telling a user you can't finish a task is better than making them wait a long time. Sometimes, you can tell them you'll get back to them and let them do something else, if the nature of the work allows that sort of thing.

Most of the design patterns affect the user interface, but not always. One of my applications maintains two caches of images, one in the cloud, one in local storage in the lab, because medical images are *huge* and it is painful to have to pull an image over a network. The cache in the cloud puts the images close to the webserver, so the application still performs adequately. The cache in the lab is used for everything else.

2.3.8 WHERE DOES VOLATILITY LIVE?

Every problem has one or more parts that will change faster than the rest, and every solution must deal with that. You can't eliminate volatility, so you do the next best thing: isolate it. You design your system so that the volatile parts that change together are as close together as possible to minimize the work required to make and distribute the changes. The result is that certain parts of your system will change nearly every release, while others change rarely, if at all.

In most applications, the most volatile parts will be in the user interface. This is a result of fickle and constantly changing tastes of users, as well as changes in the way an organization do things. You will see this most often in data-strong situations, but also in control-strong situations. Function-strong situations don't usually exhibit the same level of volatility because we don't often need to change the algorithm or processing used in one of our filters. In fact, most volatility in such situations can be addressed by moving the filters around in the pipeline.

The most useful design pattern for isolating volatility is the Layers pattern [14]. The pattern should be named "Walled Compounds" as it serves to isolate design components from each other by hiding them behind simple interfaces. The main characteristic of the pattern is the rules about inter-layer visibility and operation, rules that make the pattern so useful (*so* useful, in fact, it appears everywhere). The first rule is that a layer knows nothing about the layers that depend on it. The second rule is that a layer knows only the interface of any layers on which it depends, but nothing about the internals.

Any collection of components can be grouped into a walled compound and you can use the Façade pattern from [9] to build the interface. The Adapter and Bridge patterns from [9] also work.

Volatility is important enough as a design consideration that I developed a measure for it in the mid-1990s which we explore in Chapters 4 and 5. Chapter 4 devotes and entire section (Section 4.5) to combating volatility.

2.3.9 SECURITY, ALWAYS SECURITY

Security is a complex enough problem that a lot of people have made an entire career out of it, and there are entire government agencies focused on it. News reports about attacks by foreign parties on our cyber-infrastructure, and by us on theirs, are now daily occurrences. How is an application designer to deal with it?

Fortunately, the really big problem breaks down into two much easier problems: how do I ensure that only authorized users access the application? And, what happens if (when) these efforts fail? Let's face it, software applications are like hydraulic systems: they're going to leak. The best you can do is make it as hard as possible for a leak to occur, and as harmless as possible when one finally does.

Starting in the 1990s, convenience stores started limiting the amount of cash kept in the store, the idea being that what they don't have can't be stolen. When Target was hit with the data breach in 2012, I was working at Nordstrom which, like every other retailer, was trying to figure out how to protect credit card information. I raised the idea of using convenience stores as a model, basically not keeping credit card information on our network. I'd done some work in payment card systems at a previous employer and knew that large banks were actively encouraging their merchant customers to go this route, to the point of providing free technology to make it possible.

I now work in healthcare and protecting medical information is one of our top priorities, and one of our top technology expenses. There is no good way to operate a hospital without collecting and using medical information for its patients, so just not having it is not an option, for the moment. Changes are afoot in several industries that will make this technically possible in the near future, and maybe preferable from a business standpoint.

Most often, the pattern for authorizing users is to store user names and passwords. This has led to an entire industry of applications that manage our passwords for us. My latest application took another common route and uses a third party to authenticate users, so it no longer stores passwords.

Once a user is authenticated, we have to be able to control what they can and cannot do. The Role-Based Access [21] pattern is the most widely used because functionality can be built to work only when a user is given the role authorized to use it. Since it's easy to assign more than one role to a user, each unique piece of functionality can require its own role. Of course, the ability to grant roles to users must be tightly controlled.

Since systems will leak eventually, there are steps you can take to make the information as useless as possible when it gets out into the wild. Modern cyber-cryptography methods make protecting information at rest or in motion much less complicated and expensive than it used to be. The advent of public-private key encryption has eliminated the need for passwords in some cases. I manage the servers that run several application from my laptop which holds my private key, with each server knowing only my public key. Work is underway to extend this concept to verifiable credentials and identities of all kinds, for any kind of online transaction.

2.4   CHOOSE WISELY

Once you have a start at the patten language for your problem, it is time make some decisions about things you actually have some control over. The two main decisions are the programming style and the approach to take for the rest of the design. Both can have a significant impact on the success of the solution, and both are very hard to change without starting over.

2.4.1   PROGRAMMING STYLE

Believe it or not, your choice of programming style significantly influences the way you approach design as well the results you get. While there are many ways to structure code, those of us who've done it awhile recommend three. Each style has advantages and disadvantages and tends work better in some situations than others, but any of these styles *can* be used in any situation if you're willing to work at it.

Robert Martin, in *Clean Architecture* [1, pp. 57–91], provides a very usable summary of each style: Structured, Object-Oriented, and Functional. He also gives some advice about when to choose one style over another, focusing on characteristics of code built using each style. I take a slightly different approach.

I should make clear that this style decision is made only for the high-level structure of the code you are going to write. You may well find that you make use of all three styles as you get into the details. Feel free to do so, because the design of a method is a solution to a problem in its own right, and the needs of that method may dictate using one style over the others.

I have found that the best style of programming is partly determined by the type of problem I'm trying to solve. To be honest, my default style is object-oriented, but I don't always stick to that, and all of my applications are a mix.

Structured Programming

Structured or procedural programming was "discovered" by Edgar Dykstra in the late 1950s when he concluded that all programming consists of sequence, selection, and iteration. The main contribution of structured programming is the elimination of the *goto* statement, and especially the computed goto of FORTRAN IV.

Structured programming works best for short, single-function scripts. It works great for building the pipes in Pipes and Filters situations, and also works for building smaller filters. Prior to 1990, most applications were built as collections of functions programmed using structured programming, or worse. I once worked on a manufacturing application written in FORTRAN IV in which every event was followed by a computed goto statement. While you got used to working that way, well-structured code was something you missed.

One characteristic of structured programs is that complexity is visible directly in the code. More complex functions tend to be longer. Logic flow is also highly visible. These characteristics make structured code a bit easier to read and understand, especially when you didn't write it. Some programmers prefer structured code for this reason.

Personally, I would not create an entire application, or recommend doing so, using structured programming. You will see why in the next section.

Object-Oriented Programming

The world is object-oriented (and analog, but that's a different issue). For example, describing the motion of a fingertip in 3D space is a very complex undertaking, such as in a 3D game, until you realize the finger is attached to a hand, the hand to a wrist, the wrist to an elbow, the elbow to a shoulder, and so on. An object-oriented solution would describe the finger's motion relative to the hand and leave it at that. The hand would in turn know how it moves relative to the wrist, which knows how it moves relative to the elbow, and so on. When all of those relative motions are combined, you can get a realistic depiction of a finger's general movements.

In data-strong situations, the data model drives everything. It also provides a ready-made set of classes. These classes become models in applications based on the MVC or Application

Controller patterns. In control-strong scenarios, each type of device or sensor becomes a reasonable class in the application.

Each class is a black box with defined externally observable behavior. When designing a set of classes, the communication between classes is a major concern. Modeling classes as autonomous, message-passing entities can make understanding an object-oriented design much easier. It makes debugging one much easier, too. This approach extends to applications that communicate at the enterprise level, the only change to the mechanics are the names of the classes.

As design progresses, you can focus all of your attention on the behavior of the object at hand knowing that such behavior, when combined with other objects, will produce the desired results. One effect is that complexity moves from the individual methods into the spaces between objects; that is, it moves from the code into the design itself. Learning and following the logic in an object-oriented system can be difficult, even very difficult.

The predominant style for the backend of the case study application and pretty much all of my applications is object-oriented. The frontend follows more of a pseudo-functional style that is the habit of many JavaScript programmers.

Functional Programming

In functional code, every statement is a function call, and many statements are function calls within function calls within function calls…you get the idea. LISP is a function-oriented language. Martin's discussion of functional programming [1, pp. 83–91] focuses on the immutability of variables in functional code. Variables are initialized, but never changed, which leads to some interesting results (you'll have to read Martin's book to lean about those; you should read it anyway).

While some would argue about the functional nature of JavaScript, defining a function in the signature of another is a clear sign that functional programming is present. Granted, most JavaScript code, by volume, is structured (or not), but a lot of it is functional.

I don't think I would create an entire application using functional programming. I once built an application that had two user interfaces, one of which was built with AutoCAD's menu language which was a LISP derivative. That was a painful experience, but to each his or her own.

2.4.2   DESIGN APPROACH

When I talk about design approach, I mean the method I use to discover components that make up the system; you could call it a design strategy. Different approaches lead to very different designs, and even different architectures with different properties [22]. Sharble and Cohen [23] conducted a rather entertaining experiment by designing the same system (the control system for a brewery) using two different design approaches and got two very different designs.

There are currently four widely used approaches to design. This list was gleaned from a number of accepted methods for object-oriented design when [5] was written. Since then, additional approaches have come up, but all of them are variations on a theme, repackaging old ideas with new names. I won't cover these new approaches here because I haven't found them unique enough to add to my toolbox.

Each of the four approaches will lead to a relatively complete picture of all three dimensions (data, function, control/behavior) but they differ in the way they go about it. Each approach has its own strengths and weaknesses. Each approach is named for its primary emphasis, so you get different sets of components, be they classes, data structures, or whatever. Here's the list:

- Data-driven

- Process-driven

- Event-driven

- Responsibility-driven

Different types of problems require different approaches. Even different problems in the same space (for example, two data-strong problems) might require different approaches due to differing goals and constraints. You should become familiar with all of these approaches, even if you only ever use two or three.

We'll explore each of these in a bit more detail, but before we dive in, I will leave you this bit of advice: not one of these approaches will work in all situations. In fact, no one approach will give good results in *any* situation. Even when designing a data-strong application, I use a combination of approaches, usually data-driven and responsibility-driven. Event-driven works for the first pass through a control-strong problem, and process-driven works for the first pass through a function-strong problem. In both cases, however, I will follow up with a pass using the responsibility-driven approach.

Data-Driven

The data-driven approach focuses on the data model of the problem and is one of the two most commonly used approaches for the design of applications, especially data-strong applications. It should be very familiar to those of you who have done data modeling. The approach views components in terms of their component parts and static structural relationships with other components.

Among its main strengths is the tendency to identify the entities about which information must be retained, both active and passive (active entities perform part of the function of the application while passive entities are acted upon by others). The resulting set of entities will typically map very closely to the data model, and will be strongly reflected in a relational database structure should one be used.

Process-Driven

The process-driven approach focuses on the process model in the problem domain and works well for some function-strong situations. This approach tends to results in components that loosely align to the high-level processes in the problem domain. The set of components can easily be mapped from a set of dataflow diagrams.

The major strength of this approach is that it ensures that all processes and/or functions get included in the design. However, it may not map well to entities in the problem domain which can lead to designs with high coupling and low cohesion. Use of this approach can lead to the selection and specification of algorithms too early in the design process.

Event-Driven

The event-driven approach focuses on the events in the problem domain which the solution must detect and perhaps handle. The major strength of this approach is that it packages the solution's interaction with its environment into nice coherent components. The resulting event-response pairs can be similar to the results from the responsibility-driven approach. This approach tends to work well in control-strong situations, or when designing the external interfaces for other situations.

The major weakness is that the approach favors components that coalesce around events and responses and may not map well to the entities in the problem domain.

Responsibility-Driven

This approach focuses on assigning responsibility for actions required to fulfill requirements to various components in the application. The approach ensures that all required actions are accounted for in the design and leads to easier traceability from the design back to the requirements. The main influence in this approach is role-playing. Components in the design tend to map to groups of related responsibilities, with one being designated the *component-in-charge* of each group. This works best when using the object-oriented programming style.

This approach works well as the initial approach in control-strong situations, and makes a good approach for a second pass in other situations.

The major weakness is that this approach tends to limit focus to the current application, leading to potentially sufficient but incomplete components. Further, it can miss components, or lead to non-intuitive data structures for data-strong and function-strong applications.

2.5   WITH ALL THAT, THERE ARE SHORTCUTS

The information in this chapter will guide you toward finding the best architecture for your solution, and with it, the design approach for the rest of the project that has the best chance of success, defined as finishing the project in a reasonable amount of time with a high confidence that your solution will actually solve the problem. These various topics, however, are not independent, and if you know the natural groupings, you can make a lot of architectural decisions in very short order and be 90% of the way there, 90% of the time.

For example, if your problem is data-strong, assume that you will be using the Blackboard pattern with a shared data store and the data-driven approach to design (I always make a second pass using a responsibility-driven approach). You may decide to use object-oriented programming as part of the package. These three ideas go well together.

If your problem is function-strong, assume you will be using the Pipes and Filters pattern and primarily a responsibility-driven approach to design. You may still opt to use object-oriented programming, but either of the other two styles, structured or functional, work quite well.

If your problem is control-strong, start with the Microkernel pattern and the object-oriented programming method. Use an event-driven approach to design, followed with a responsibility-driven approach.

Finally, if you plan to host your application on the web, even though it runs on a single computer, plan to use the Layers/Walled Compounds and Application Controller patterns; there really is no good way to avoid them.

After a few years, you will be able to have a five minute conversation with someone about a problem they're having and instinctively make most of these decisions. You will find that with practice, you will rarely have to change your plans as work progresses. That said, stuff happens, so be on the lookout for some reason the usual combination of patterns and approaches might not work. It doesn't happen often, but build enough systems and you will encounter one.

2.6   SUMMARY

This is the shortest treatise on software architecture you will ever read. The subject is huge and complex and many much larger books have been written on the topic. Most of the literature on software architecture concerns how to model and document the architecture. Here, I've focused on what needs to be done to ensure that you create a solution that can actually solve your problem. The techniques, patterns, and advice in this chapter were honed by long experience and repeated successes. They work. They don't work in every case, but when you encounter such a case, you can at least figure out how to modify the advice to fit your situation.

Now, let's move on to the first part of designing the details of your solution: describing the problem, the second-most important step.

2.7   FURTHER READING

It is no coincidence that most of my preferred sources on architecture are about patterns and pattern languages. By studying the pattern languages used by others, you can learn a lot about the craft of designing software architectures. Some of my favorites, in addition to those appearing elsewhere in this chapter, are:

- Alexander, C., Ishikawa, S., Silverstein, M., Jacobson, M., Fiksdahl-King, I., and Angel, S. (1977). *A Pattern Language: Towns, Buildings, Construction.* New York, Oxford University Press.

- Schmidt, D., Stal, M., Rohnert, H., and Buschmann, F. (2000). *Pattern-Oriented Software Architecture: Patterns for Concurrent and Networked Objects*, vol. 2. New York, John Wiliey & Sons.

- Kircher, M. and Jain, P. (2004). *Pattern-Oriented Software Architecture: Patterns for Resource Management*, vol. 3. Hoboken, NJ, John Wiley & Sons.

- Buschmann, F., Henney, K., and Schmidt, D. C. (2007). *Pattern-Oriented Software Architecture: A Pattern Language for Distributed Computing*, vol. 4. Hoboken, NJ, John Wiliey & Sons.

- Buschmann, F., Henney, K., and Schmidt, D. C. (2007). *Pattern-Oriented Software Architecture: On Patterns and Pattern Languages*, vol. 5. Hoboken, NJ, John Wiley & Sons.

- Bass, L., Clements, P., and Kazman, R. (2007). *Software Architecture in Practice.* New York, Addison-Wesley.

- Rosanski, N. and Woods, E. (2012). *Software Systems Architecture: Working with Stakeholders Using Viewpoints and Perspectives.* Upper Saddle River, NJ, Addison-Wesley.

- Coad, P., North, D. and Mayfield, M. (1997). *Object Models: Strategies, Patterns, and Applications.* Upper Saddle River, NJ, Yourdon Press.

- Fowler, M. (1997). *Analysis Patterns: Reusable Object Models.* Reading, MA, Addison-Wesley.

In addition to sources already mentioned, I relied on the following as I was learning to *really* write code:

- Jacobson, I., Christerson, M., Patrik, J., and Övergaard, G. (1992). *Object-Oriented Software Engineering: A Use Case Driven Approach.* Reading, MA, Addison-Wesley.

- McMenamin, S. M. and Palmer, J. F. (1984). *Essential Systems Analysis.* Englewood Cliffs, NJ, Yourdon Press.

- DeMarco, T. (1978). *Structured Analysis and Specification.* New York, Yourdon, Inc.

- Page-Jones, M. (1980). *The Practical Guide to Structured Systems Design.* Englewood Cliffs, NJ, Yourdon Press.

These two texts are among my primary sources on approaches for design, as well as two of my favorite sources on object-oriented software development:

- Embley, D. W., Kurtz, B. D., and Woodfield, S. N. (1992). *Object-Oriented Systems Analysis: A Model-Driven Approach.* Englewood Cliffs, NJ, Yourdon Press.

- Wirfs-Brock, R., Wilkerson, B., and Wiener, L. (1990). *Designing Object-Oriented Software.* Englewood Cliffs, NJ, Prentice Hall.

CHAPTER 3

Solve the Right Problem

Designing software is not very different from designing any other complex structure: Few people are good at it; no single recipe always produces a good product; and the more people involved, the smaller the probability of success.
— Maarten Boasson.

In the last chapter, we learned that the architecture is driven mostly by non-functional requirements. Functional requirements do matter and we focus on them in this chapter.

There is a general feeling design is bad, that its main purpose is to fill notebooks with diagrams that will never be used. This feeling is particularly prevalent among Agile developers. What they really mean is that design as an explicit phase is bad, and I agree. They do *not* mean that design should not be done; it is impossible to create software without it. In a classic waterfall project, the Design phase is when every structure and function gets designed to the lowest level of detail before any code gets written, which has never been possible.

Reality is somewhere in the middle. The main abstractions and responsibilities need to be defined and designed to the extent we know how they work into the rest of the design. Interfaces between major abstractions have to be worked out in advance or you risk having to rework large parts of the code multiple times. Rework is waste, even if you call it refactoring.

The last chapter led you through developing your architecture. While the architecture gives you an idea of the major pieces, it doesn't provide any details about the components you need to solve the problem. In this chapter, we define the problem in detail, interpreting the story problem to determine the equation to be solved. This is not a trivial matter since you can easily set out to solve the wrong problem, and once down such a path, there is no hope for correction; you have to start over. This may sound a lot like analysis, and it is, but there is a fair amount of design in this process, so I prefer the term *problem domain design*. The primary purpose of this activity is to identify the essential abstractions in the problem domain. An abstraction is an element in the domain model which represents all or part of a concrete or conceptual "thing" in the problem domain [24].

The next chapter describes the process of defining the solution to the problem, itself a difficult journey fraught with misadventure and danger.

On nearly every project, I have encountered two issues time and again, neither of which is technical. The first is whether the design is correct, whether I have found *all* of the components I need to solve the problem. I also worry about whether I have included *only* those components needed to solve this particular problem. I've never been confident that any of the design ap-proaches described in Chapter 2 can guarantee great results as they all have significant blind spots.

The second issue is that I *really* dislike having to rewrite large sections of code to change a set of interfaces, especially when the work could have been avoided by thinking about the problem for another 10 minutes. Designing component or function interfaces is an oft-neglected activity in the design of any solution. To combat this, I have included specific activities to make sure that a component or function has access to all of the data it will need. In my C++ projects, I took to explicitly assigning the responsibility to create and delete objects; chase down enough bad pointers and you can see why this is the easy way out.

As an aside, you may find that I wander off into the terminology of object-oriented design and development. I've been using object-oriented techniques for well over 30 years and I find that it matches rather well the structure of the world. Others disagree, and that's ok. I will try to use the term *abstraction* rather than classes throughout this book. An abstraction, after all, is nothing more than a collection of properties—attributes and behaviors—possessed by the domain "thing" it represents. If that sounds a lot like the definition of a class, it is and I consider the two terms interchangeable. When you read "classes," don't interpret that to mean whatever I'm saying works only in object-oriented development. Translate "class" to whatever term describes a collection of properties in your mind.

3.1   FOUR DOMAINS OF DESIGN

I've been around long enough to know when an idea is good enough to steal (with full attribution to the original source, of course). Such is the case with a paper written by George Yuan [25]. Yuan described a full analysis and design process for object-oriented software, but that isn't the interesting part. What I found fascinating was the way he split the problem and solution into four domains. Until that time, Yuan was the only person other than me to talk in terms of a creating a model of the problem domain. I incorporated his thinking into my own work nearly 25 years ago and haven't looked back. I highly recommend this particular idea.

Yuan divides the problem and the solution into four domains: Problem Domain, Ap-plication Domain, Application-Specific Domain, and Application-Generic Domain, shown in Figure 3.1. We'll briefly explore each as they are important to the design process. Yuan writes in terms of objects, but we'll go with *abstractions* because they won't become objects unless and until we decide to use object-oriented programming. Regardless of our choice, we will still need to deal with these abstractions in some way.

It is important to note that the concerns in each domain are specific to the problem at hand. The application domain abstractions for an operating system may become some of the application-generic abstractions for an application written to run on that operating system.

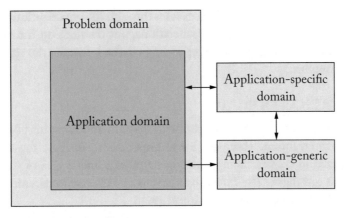

Figure 3.1: The four domains of design.

3.1.1   PROBLEM DOMAIN

The problem domain contains abstractions that appear in the real world, from the perspective of the user and system requirements, that may be part of or interact with the application. These abstractions represent people, organizations, things, ideas, and rules that you need to consider even if you don't include them in the application. The problem domain describes the problem structure and business logic, saying nothing about the implementation. Abstractions in this domain relate to each other through three basic structures.

1. *Generalization/specialization (is-a)*: A relationship in which one abstraction is a specialization or subtype of another. The specialization depends on its more general "parent" (a common term for the generalization). It is a directed relationship from the specialization to the generalization. All properties and behaviors in the generalization are present in the specialization. The structure is often shortened to "gen-spec" in conversation.

2. *Association (uses)*: An association exists when an instance of one abstraction is structurally related to an instance of another, such as a customer to an order. In some cases, such as an order, the association has its own attributes and behaviors and becomes an abstraction in its own right. An association has direction and cardinality: one-to-one (1:1), one-to-many (1:m), or many-to-many (m:n).

3. *Aggregation (has)*: An aggregation is a special case of association and designates that one abstraction fully contains one or more other abstractions, such as the way an assembly contains other assemblies or individual parts.

A sales order in a wholesale business is technically an association between a customer, a shipping location, and one or more items. Attributes include item price, item cost, the terms and conditions of the order, and one or more shipping dates. Behaviors are the various operations that cause the state of the order to change as it progresses through the organization. The organizations participating in the order are usually pure associations, but the lines on the order that relate to the items may appear as an aggregation. Depending on the business, the structure of an order can become quite complex.

3.1.2 APPLICATION DOMAIN

The application domain is a subset of the problem domain and contains abstractions that will be implemented in the solution. There is a lot to unpack here, because the decision to include or exclude an abstraction creates the application boundary, and is one of the most important design decisions you can make. The more abstractions you choose to include in the application domain, the more abstractions you need to implement.

Some problem domain abstractions are part of the data model and need to be included along with their relationships in applications of all three main types. Other abstractions are not part of the data model but directly interact with abstractions that are and need to be dealt with. These will likely have a limited representation inside the application, such as groups of users or external organizations that provide data.

Other abstractions in the problem domain may represent business rules or business logic and may need to be part of the business logic in the solution. In function-strong problems, these abstractions may represent mathematical operations or models that need to be applied to data; these should be included in the application domain. They may not become objects, but even if you implement objects, they will be included in the behaviors of some objects.

In a control-strong problem, the events or devices of concern to the application may be abstractions in the problem domain. Unless you need to model the devices directly, to monitor maintenance for example, you might not need to include the devices themselves in the application domain, but will need to model the events and create an interface so the device can send data into the application, and you will need an application domain abstraction to receive that data.

3.1.3 APPLICATION-SPECIFIC DOMAIN

The application-specific domain includes abstractions crafted specifically to support this particular application. These support abstractions may be grouped into subdomains such as user interface, data access, authorization, or communication, and often run vertically through the application.

In our case study application, built using the Application Controller pattern, this domain includes HTML templates, JavaScript functions, and Django views.

3.1.4 APPLICATION-GENERIC DOMAIN

The application-generic domain includes support abstractions that can benefit multiple applications. Code libraries fall into this domain, and may implement functions such as logging, database management, messaging, and other generic functions. I have used libraries for machine learning, graph cycle detection, numeric processing, and many other functions.

The case study includes additional functions built for the application that have since been used in other applications, including the implementation of audit trails and facilities for changes that require user review and approval.

3.2 PROBLEM DOMAIN DESIGN

Once the architectural decisions have been made, the next step is to define the problem. Here, you take the basic shape of the problem identified in the work of Chapter 2 and bulk it up with specific functionality. The problem domain model takes shape during this activity which forms the starting point for the rest of the design. The abstractions in the model represent those concepts that are either central to your application or directly connected with it. Both will materially affect your solution. To do this, you identify relevant abstractions and assign and allocate to them roles, responsibilities, attributes, and functions. You do this from a business perspective, keeping the business problem front and center.

The goal is to solve the problem effectively and correctly; to identify the set of abstractions that will solve the problem without too much wasted effort. This is the divining of the equation out of the story problem; if you're not careful, you can easily end up solving the wrong problem.

The problem domain does not often provide obvious choices about where functionality should go and there is usually some sort of tradeoff. The boundary around the application hasn't been determined yet, so feel free to explore alternative boundaries. You will find that you need to adjust and readjust the tradeoff decisions to meet the goals and constraints of the project.

Three main design concepts come to bear on this activity, which together form what I call the *quality of abstraction*:

1. Sufficiency [5, pp. 368–376]—Does the model contain enough real-world facts and concepts to implement the application?

2. Completeness [5, pp. 377–383]—Does the model contain enough real-world facts and concepts to be generally useful to the wider enterprise or the market at large? Note that a more complete model makes future modifications much easier.

3. Cohesion [5, pp. 384–393]—Does each abstraction represent one and only one (or part of one and only one) real-world abstraction, including all of it roles, responsibilities, and expectations? Cohesion is the notion that a problem domain abstraction should represent one and only one real-world entity, concept, person, organization, or thing. A simple way to measure cohesion at this level is to look at the properties of your abstraction: does it contain properties from more than one abstraction in the real world? If the answer is yes, your abstraction is not cohesive.

This activity consists of a number of tasks, each the result of having to stop work and figure something out during a project. For the record, this is *not* a methodology. These tasks provide information you will need later. While you will generally do them in the order given, you can skip around or come back to a task at any time. However, skipping any of them altogether will cause you grief down the road.

These tasks don't take as long as they seem. You already know enough about the problem to sketch out the abstractions that matter, and you have some idea about what they need to know and do. Don't worry about details such as data maintenance since basic functions like that exist in any system that retains data, and makes up most of the work in data-strong situations. You will get to that later; for now, focus on the business-visible functions that matter to the problem at hand.

Remember, the focus of these steps is to identify what *already exists* in the problem domain. You are not making things up or adding "features." Your sole purpose is to describe your problem.

3.2.1 IDENTIFY SYSTEM RESPONSIBILITIES

The responsibilities of a system include functions it must perform, events to which it must respond, products it must produce, and assistance it must provide to business processes. Taken together, the set of system responsibilities is the set of functional requirements for the application.

Different development methods contain varying expectations for the amount of time to spend identifying these responsibilities and the degree of formality used to document them. Whatever works for you is good enough for our purposes. While a lot has been written on ways to gather and document requirements, none work in all cases, and are overkill in most. Remember your audience for requirements: the developers who need to build the system (you will get a lot of argument about this from business analysts, so be ready). I find that a data model, a set of use cases and products (if any), and some wireframes for the user interface are usually sufficient for most data-strong problems. A set of processing steps or algorithms are most of what you need for function-strong problems. Control-strong problems require a rather complete event model along with any serious timing or concurrency constraints.

Identify groups of users, customers, and others who have a stake in the solution. Classify them according to their interests, skills, and potential frequency of use. These groups help populate the project communication plan, but also go a long ways toward defining the roles you will need if you decide to implement tole-based access control.

Identify external events not originated by users to which the system must respond, as well as the system's required response. Include data feeds to and from outside systems. These may lead to additional system responsibilities, especially in a control-strong situation.

Look for coherent activities that cut across the application. When performing a task, a person might take one of several paths through the work, forming a workflow. Look also for workflows that exist outside of your current set of system responsibilities, and especially for workflows that might intersect with the system. These workflows point to use cases that likely need to become system responsibilities. If you find new responsibilities, add them to your set.

You may notice that many of the following steps have you iterate through the system responsibilities and do something for each one. There is no harm in doing each step as a system responsibility is identified. After you have addressed all of the system responsibilities, you may find that you have duplicate abstractions, structures, communication links, sets of attributes, services, or message connections. This is good news, the duplication tells you that what you have found participates in the application more than one way. You can eliminate the duplicates.

In the case study, the main purpose of the application is to assist in the creation of sales and purchase orders, manage inventory to allow an overall reduction in the level of required inventory and enhance the productivity of the sales, purchase, and warehouse people. The system is the primary application for the entire division, assisting all front- and back-office functions that directly participate in the business of the division. Support functions such as Finance and Human Resources use separate systems with which the case study interfaces.

The system responsibilities in our case study include the following.

1. Enable the entry and tracking of sales orders supporting standard industry practices (the lumber business is a little weird), including release to the warehouse for picking and shipping.

2. Enable the entry and tracking of purchase orders based on available and projected inventory, including support for practices unique to the lumber business (it is common practice to order a year's output from a lumber mill and mark the purchase as TBD because the mix and amount of product, the cost, and the ship dates are all unknown, that is, to be determined. The mill notifies the purchaser when a shipment is about to be sent, including the mix of product).

3. Manage inventory in a way that reduces the overall level required to support business activity, including the ability to source sales orders from incoming purchase receipts.

4. Support common industry practices for pricing, including full-truck and full-train car load discounts, as well as discounts for receiving random length product (the buyer orders a quantity of a type of board, say Douglas Fir kiln dried 2x4, in terms of total board feet of lumber, but doesn't care what lengths are received. A board foot is 1" x 12" x 12"—an 8-foot 1x12 is 8 board feet—and is the standard unit of measure for lumber worldwide). Side note: units of measure are a huge pain in this business.

5. Support industry standards for product substitution and back orders (out of stock items on an order are usually just canceled, although back orders are still fairly common).

6. Enable the recording and tracking of work orders to customize some types of products prior to shipping, mainly cutting architectural and engineering beams to the correct length.

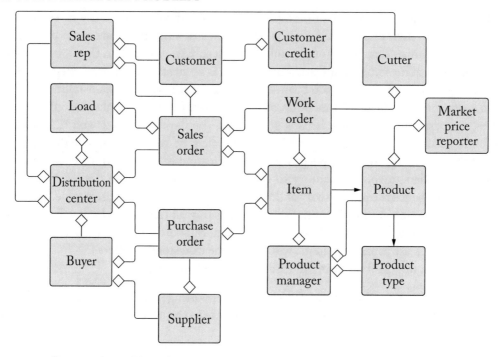

Figure 3.2: Case study problem domain model.

7. Enable the planning and optimization of truck loads on outbound shipments, subject to the constraints of load weight and volume restrictions.

8. Manage customer credit to prevent sales orders from being released unless sufficient credit is available (this presents a number of very interesting problems).

9. Capture activity across all remote locations and submit invoices to finance on a thrice-daily schedule.

10. Provide for business intelligence and data analysis activities without affecting production work.

Between the previous paragraphs and the model in Figure 3.2, you get a pretty clear picture about the overall set of requirements.

3.2.2   IDENTIFY PROBLEM DOMAIN ABSTRACTIONS

Choosing the abstractions to include in the problem domain is one of the most critical activities when building an application. I use *choosing* instead of *finding* because the decision to include an abstraction might come down to a tradeoff, and as the designer you often have considerable latitude about how to make that trade. This is particularly true when modeling relationships between abstractions. An aggregation, for example, can look like a Bill of Materials, which is a recursive tree-like structure (you will encounter this type of structure a lot), or a list of ingredients in which the individual components are not recognizable in the aggregate. To tell which of the two situations you face, ask if the components are still recognizable when they are removed from the aggregated abstraction.

Your two main concerns are as follows.

1. Have you identified *all* of the abstractions from the problem domain you need to build the application?

2. Have you identified *only* those abstractions you need to build the application?

The first question deals with sufficiency while the second deals with efficiency; both deal with completeness. At first glance, the questions appear to be contradictory, with the first seeking more abstractions than you might need for the second. They are not in conflict if the questions are applied in the order given, so you first identify abstractions in the problem domain that *might* be needed, then from those select the abstractions you *actually* need. When applying the second question, if you find some abstractions that don't look like they will be required or even referred to by the application, you can safely remove them.

The process is fairly simple, and can be done quickly: for each system responsibility, identify the real-world entities, concepts, ideas, policies, organizations, people, events, and processes that contribute in some way to completing the responsibility. Each of these "things" will become an abstraction in your model. After you work through all of the system responsibilities, you will have several sets of abstractions that will include many repeated abstractions. Your problem domain model is the union of these sets (eliminate duplicates).

Since the case study is a data-strong application, most of the problem domain abstractions will appear in the application domain. The problem domain model for our case study is shown in Figure 3.2. The model is simplified to include only the main concepts and entities in the problem. We will add many others in the application domain in order to deal with some of the peculiarities of the lumber business. These new abstractions go in the application domain because they represent one way to solve a given problem. The design decisions were open for debate, and one in particular generated a lot of it.

A quick check over the responsibilities in the previous section tells us that we can assign them to one or more of the abstractions already identified so we don't need to add any. This is often the case with data-strong problems, but rarely the case for function-strong and especially control-strong situations. In any case, it's a step that needs to be done lest we miss something.

3.2.3 IDENTIFY STRUCTURES

Abstractions do not work alone; they work together, sometimes as static structures, sometimes dynamically. The purpose of this step is to find the static structures that exist in the problem domain, using business rules as the basis for our search. Sometimes, we need to add an abstraction to represent the structure itself. For now, focus on static structures, those that exist no matter what, we'll cover dynamic relationships in the next section.

Abstractions form two basic kinds of static structures [10, 25–27]. *Generalization-specializations* (gen-spec) are structures in which one abstraction, the specialization, extends the properties of the other, the generalization. You can think of this as a form of subtyping. While we haven't identified properties yet, familiarity with the business domain should help you identify the gen-spec structures that matter most. You can add others as you find them.

An example of a specialization would be a particular model of a sport-utility vehicle (SUV), the generalization for which would be all types of SUVs (which is itself a specialization of passenger vehicles). There are times when all that matters is that a vehicle is an SUV; the make or model do not matter. There are times, however, when a particular model of SUV has capabilities that others don't, and the specialization is required.

In most designs, generalizations are added as part of the design process, although you sometimes find gen-spec structures that natively exist in the problem domain. When you do, they are always important. In our case, Item is a specialization of Product, which is a specialization of Product Type. This hierarchy is important for a number of reasons. An item might be a Douglas Fir kiln-dried 8-foot 2x4; the product would be Douglas Fir kiln-dried 2x4; the product type would be Douglas Fir kiln-dried structural lumber (green lumber is another product type). Each product type has a unique set of attributes, a problem we'll have to solve in the application design domain.

In many situations, these hierarchies define the universe of substitutions. In the case study, an item can be substituted for another of the same product if it can be modified for the same use (a 12-foot 2x4 can be substituted for an 8-foot 2x4 by cutting it; the reverse is not possible). Some cross-product substitutions are allowed, such as using Western Hemlock in place of Douglas Fir (there is actually a product type where the species is Hem-Fir which means that you could get a mix of both species).

Pricing is done at the product level and is expressed in terms of price per board foot. Lumber prices are set at the mill by the market and are published twice weekly. Prices are by species, shape, grade, and one or two other categories. Publishing prices is a good-sized business all by itself. Determining the correct price is a function of the Product, so we leave that out of the problem domain for now.

Aggregation are whole-part structures, where one abstraction (the aggregation) fully contains another (the component), or where one abstraction "supervises" another. Aggregations are very common in the world, The most common example of an aggregation is an assembly built from component parts. Buildings, machines, and vehicles are all examples of assemblies. Beams, windows, doors, engines, and such are all examples of components. A software application is an assembly of software and other types of components. While clearly visible in the problem domain, there are a number of ways to implement aggregations in the detail design, decisions we will happily put off until later.

3.2.4   IDENTIFY COMMUNICATION PATTERNS

Often, two or more abstractions must work together dynamically to fulfill a system responsibility; they must communicate. What and how they communicate is left for later design work; for now, it is enough to know that they communicate in some way.

Work each system responsibility one at a time, identifying these communication links. The result is a network of communication links for that system responsibility, represented as directed connections between pairs of abstractions. Across all system responsibilities, the union of these subnetworks form the network of links for the application (a link between a specific pair of abstractions needs to be counted only once).

The density of these communication links, in terms of the ratio of links to abstractions, is a measure of complexity [5, pp. 330–350] in the problem domain. There is nothing you can do about this complexity, you can only add more, and you will, as you continue to design.

Jacobson's use cases [28] are an excellent tool for identifying and documenting the subnetworks for each system responsibility.

3.2.5   IDENTIFY TIMING AND CONCURRENCY CONSTRAINTS

Sometimes, you run into timing and concurrency constraints that are inherently part of the problem domain, especially in control-strong situations. These constraints can lead to new abstractions, attributes, and methods which need to be added to your model.

A timing constraint exists when the system must respond to an event within a certain amount of time. You also find these kinds of constraints in inventory systems in which the goods tracked are perishable, such as drugs, medical tissue samples, or food. Timing constraints make the design process more complex, meaning it will be more difficult and take longer and can change your approach to the rest of the design, or even where you plan to host the solution.

Concurrency constraints exist when operations or processes must be done in sequence, rather than in parallel. Problems that allow for more parallel operation give you much more flexibility, especially when figuring out ways to increase the scalability of the application.

For now, just note these constraints in your model; you will deal with them later.

3.2.6   IDENTIFY AND DESCRIBE ATTRIBUTES

For each abstraction in the problem domain, name the descriptive properties (as opposed to behaviors, those come later) that need to be exposed in the application. These properties create your initial set of attributes for each abstraction. I want to emphasize that this is your *first cut* at the sets of attributes; you *will* add more and may be able to remove a few, but don't count on it.

As you discover attributes, define them in terms of a descriptive name, their type (string, number, date-time, or Boolean value), and the domain. An attribute's domain is the set of allowable values the attribute can take. If you can, create sets of selection options for your string attributes as plain strings can create security problems. Also note any constraints that apply to an attribute as they will come in handy later. Do this as soon as you add an attribute to your model; it is much easier to do it now than later (you'll have to trust me on this, or don't, and suffer the consequences yourself).

In practice, I create the most obvious and obviously necessary attributes in my initial pass through the problem domain. These can be gleaned from the way people talk about the abstractions. One of the big design arguments in our case study was about addresses (see Section 4.9 for a more detailed discussion). Fundamentally, an address points to a location and people normally associate a customer with one address. When dealing with large chains, however, there may be up to three addresses for a "customer" on a sales order (the relationship between Customer and Sales Order is complex and rather fluid):

- the address where the bill is sent (which may be a different customer than the one receiving the goods),

- the address where the "maker of the deal" is located (which may also be a different customer than the one receiving the goods), and

- the address where the goods will be delivered.

This has implications for how we represent an address in the data model, but it is clear that "address" is *not* an attribute of Customer. The same goes for Supplier. In contrast, the address structure of a Distribution Center is relatively simple. Customers also have a credit balance, but even that can be more than an single value.

There are many other obvious attributes you can note and include in your model. Here, though, I usually focus on those that provide clues about the business rules and relationships, as they will influence other parts of the model. "Regular" attributes that are solely descriptive in nature can be safely added later.

3.2.7   IDENTIFY INSTANCE CONNECTIONS

Look for pairs of abstractions for which individual instances can be linked to identify associations. The business vocabulary provides lots of information about associations. The relationship between customer, item, and location, usually identified as a sales order because it has its own attributes, is an obvious example.

An association can be one-to-one, one-to-many, or many-to-many. If an association has attributes or behaviors of its own (such as the aforementioned order), create an abstraction to contain them. Don't worry about many-to-many associations just yet. While it is common practice to replace them with a new abstraction and a pair of one-to-many associations, we'll do that during the application domain design in the next chapter (see Section 4.10).

In data-strong problems, most of the relationships, by count, are connections between instances of domain abstractions, rather than structural relationships.

3.2.8 IDENTIFY ABSTRACTION STATES

The states or statuses through which an abstraction progresses in its lifecycle are part of the problem domain, and many of those lifecycles need to be reflected in the application. Information about the state of an abstraction comes from a number of sources.

Some attributes exist only to record the current state and are called *state attributes*; they can be safely removed from the model unless they are a legitimate part of the problem domain (but even then, consider their removal). You may add them back later to quickly identify the state of an instance, but you will also need to maintain them, which creates extra work and more sources of errors.

The values of some attributes might be constrained by the current values of other attributes. In these cases, there is a state corresponding to each of the possible values of those controlling attributes, or combinations of values when there is more than one. In other cases, the value of an attribute is controlled by other circumstances. Each of these circumstances defines a state.

A message received by an abstraction is an external event. If any attributes change values as a result of the response to this message, two states have been defined, one before the response, and one after.

The collection of states and the events that lead an abstraction from one state to another, called *transitions*, form a state model for each abstraction. The size of this set, and the density of transitions (the transition count divided by the state count) is another form of complexity inherent in the problem domain.

While the state models provide great information, they are not always important, but can be very important. In a wholesale business, such as our case study, the state model of a sales or purchase order can reflect the entire structure of business operations, in turn shaping the entire application.

3.2.9 IDENTIFY SERVICES

This is where things get interesting. To fulfill a system responsibility, participating abstractions must provide one or more services to other abstractions. These services may or may not evolve into actual methods (for objects) or functions (for everything else), but we'll work that when we get there.

Every system responsibility is assigned to an abstraction to oversee and coordinate its fulfillment. This abstraction is the *abstraction-in-charge*. If the system responsibility is to respond to an outside event, the abstraction-in-charge is the one that detects the event. This isn't as silly as it sounds. Explicitly assigning a system responsibility to a specific abstraction makes it much easier to design, build, test, and modify the rest of the application. It's the design equivalent of "Who's in charge here?" which is usually the starting point for any sort of troubleshooting later on.

This isn't necessarily easy, either. One of the responsibilities of the case study is to record the shipping of a sales order as a revenue event, and the receipt of a purchase order as a cost event. In both cases, notification needs to be sent to the General Ledger. These events turn on the transfer of ownership of the inventory and represent significant accounting events in the business. Shipping an order on the last day of the fiscal month helps the current month's sales and inventory figures if the event is recorded then, while delaying the recording until the next weekday might mean the event and its accounting record occur in different time periods. In a large business, this can be a significant problem.

One way to address this is to assign the responsibility of notifying the general ledger of a sale to the sales order as a result of transitioning from loaded (on the truck) to shipped. Likewise, the purchase order is assigned the responsibility to notify the general ledger when it transitions from in-transit to received. There are several other actions in the case study that trigger general ledger events, and a good practice is to assign the responsibility of doing so to the service that causes the transition.

These services must be defined, or at the very least, described. Fully defining a service isn't hard, and the information in the definition will be used to build and test any implementation of the service. Over many years, I have evolved a list of information I need to determine before I can implement a service. Each of these items has to be determined at some point, and I find it easiest and safest to do it up front. While I don't advocate binders full of design specs, I do recommend you write these down as you determine them. The resulting document is not nearly as important as the thinking that went into it, but the document will prove useful.

The following items are critical to defining a service.

1. Name the service. This should be a short (three to five word) verb-phrase that gives a clear idea of what the service does. If you can't name a service without using "mushy" words like "process" or "manage," you don't understand it well enough, or you need to split it up. Example: Generate Invoice is a service of Sales Order.

2. Describe the service in two or three sentences.

3. Identify preconditions—conditions that must be true in the problem domain before the service can execute. This is more important than you may realize. Example: a sales order cannot be released to the warehouse unless there is sufficient customer credit to cover the entire order.

4. Identify postconditions—conditions you can assume to be true after the service has completed. Example: After Generate Invoice on a Sales Order, there is an entry in the General Ledger for the event and inventory has been reduced for the items on the order.

5. Identify those conditions that do not or cannot change that might influence how this service works; these become invariant conditions, or just invariants. An invariant may appear as both a precondition and postcondition, or may be a condition that *cannot be false* in a

stable state, although some invariants can be violated while transitioning from one stable state to another.

6. List the states in which this service may be requested. These should be covered in your preconditions, but this alternate point of view can uncover new conditions.

7. Identify external attributes the service needs in order to work. Ignore attributes that belong to the abstraction to which the service belongs; you're looking for information that is not inside the abstraction.

8. Identify attributes that are modified or need to be returned after the service completes. Example: available customer credit after a sales order is released, or an inventory balance after an invoice is generated.

9. Identify the services required of other objects. These should be visible in the communications network you created in Section 3.2.4. If you find you need to add a service, do so, and add the corresponding communication link.

10. If the domain contains any constraints or rules about how the service is to be done, note them now. These might include calculations that are defined by regulations or others.

Preconditions and invariants describe the environment that must exist before a service can be used. Postconditions and other services used describe the environment after the service has completed. These two environments form the basis for black-box test cases that can be run during a review or while running code and systems tests after you have running code. For now, compare these test cases against the requirements to make sure you haven't missed anything.

The attributes in items 7 and 8 describe the external data the service will need or will provide when it runs. They could become parameters in a function or method call, or you can implement it some other way. We'll decide what to do in the next chapter.

3.2.10   IDENTIFY MESSAGE CONNECTIONS

A message connection is created when a service of one abstraction calls a service in another abstraction. Message connections are one-way links between abstractions, and should have been identified as part of the work in step 9. Message connections are the third main type of association between abstractions and like the others create a form of coupling [5, pp. 351–367]. In this case, the calling service (and its abstraction) depends upon the called service (and its abstraction). One of your objectives as an engineer is to minimize "bad" types of coupling, but here, you are identifying the coupling inherent in the problem domain. You can't eliminate it, but you can make it as harmless as possible. We'll see how in the next chapter.

3.2.11   SIMPLIFY THE MODEL

When you get to this point, you will look and your model and mutter an expletive or two. Trust me. Inevitably, your model will be a complex mess of abstractions, relationships, and messages. It's time to simplify, and shift from analysis mode to design mode. When I first encountered the project that built the case study there were over 750 entities in the data model (application domain, so it included entities added to support implementation). I was able to reduce that to just over 400 before we started construction (the wholesale lumber business is nothing if not complicated).

You may notice that some of your abstractions are similar to others. There are four kinds of similarity [5, pp. 401–413], but only three matter at this point (the fourth, functional similarity, will come up later):

- Structural: Two abstractions can have a similar set of attributes, services, or both. Services can be similar if they require similar inputs, produce similar outputs, or use similar set of services from other abstractions. If two abstractions are similar enough, consider factoring out the common parts and forming a pair of gen-spec relationships (see Section 4.6). Sure, you add a new abstraction, but you can cut down on a significant number of duplicate attributes or message connections. Fewer of either is always a good thing.

- Behavioral: Two abstractions are behaviorally similar if they have similar sets of states, respond to similar sets of events, or have similar responses to events. Like with structure, you may be able to factor out the similar behavior into a generalization that adds an abstraction but cuts down the duplicate event responses or states. It's usually a tough call, but can be rather obvious.

- Purpose (or semantic): Two abstractions are similar in purpose if they exist for the same reason. In our problem domain, two abstractions exist for the same reason when they represent the same problem domain concept. While it is rare for two abstractions to overlap completely, it does happen. More often, though, the two abstractions only partially overlap and it may make sense to combine them into a single abstraction, bringing along the attributes, relationships, and connections of both.

In general, you can reduce the overall complexity of your model by combining abstractions into gen-spec or aggregation relationships at the cost of increasing the number of abstractions. This is one of the first tradeoffs you will encounter. Your task at this point is to optimize the overall complexity of your model. Note I didn't say "minimize" because while you can do that, it usually doesn't pay off in the long run. The level of complexity you end up with is your new floor to which you will add as you progress through the rest of the project.

3.3   SUMMARY

You may feel, when working on a data-strong problem, that you are pretty much done with the problem domain design once you have the domain data model. Not so fast. Unlike architecture, in which various types of patterns tend to go together, every problem domain has something that doesn't quite fit. The steps in this chapter are like aviation regulations: a crash occurs, investigators figure out what happened, and new rules are created to prevent it from happening again. Like most endeavors, you learn more from your mistakes in software development than from your successes. Get a head start and learn from mine; it's always less painful to learn from someone else's mistakes.

You may have noticed that I use an object-oriented approach to defining the problem domain or that all of my references are about object-oriented analysis and design. To quote Robert Martin, a class is "simply a coupled grouping of functions and data" [1, p. 93]. Technically, a class is a particular implementation method for an *abstract data type* but "class" is easier to type (and doesn't require yet another acronym). The world is a lot easier to understand and model using object-oriented techniques. Just because you think about design using objects doesn't mean you develop using them, so don't let my use of objects and classes upset you.

In this chapter, we created a model of our problem domain, identifying the relevant abstractions, structures, services, and links for our application. Yet all we have are the raw materials out of which we must build our software. We start that in the next chapter.

3.4   FURTHER READING

This is perhaps the most influential book on software requirements I've encountered:

- Gause, D. C. and Weinberg, G. M. (1989). *Exploring Requirements: Quality Before Design*. New York, Dorset House.

Some additional resources on requirements:

- Davis, A. M. (1990). *Software Requirements: Analysis and Specification*. Englewood Cliffs, NJ, Prentice Hall.

- Wiegers, K. E. (2003). *Software Requirements*. Redmond, WA, Microsoft Press.

- Miller, R. E. (2009). *The Quest for Software Requirements*. Milwaukee, WI, Maven Mark Books.

CHAPTER 4

Engineer Deliberately

Software—the stuff of computer programs—is composed of sequence, selection, iteration, and indirection. Nothing more. Nothing less.

–Robert C. Martin

By now, you should have a pretty good handle on the problem you are trying to solve, and some idea of how you plan to go about it. Your problem domain abstractions are probably aligned well with abstractions in the real world, but may be too unwieldy to implement as they are. In particular, every application needs to have a user interface, business logic, and data access components, even if they don't necessarily align with the problem domain abstractions.

Your problem domain design provides a couple of minimums in terms of size and complexity. You can only add to these as you engineer your solution, so approach the rest of the project as an optimization problem. You will trade increased size and complexity for making your solution easier to build, use, deploy, or change, or all four if you're lucky. Make sure these are deliberate trades, made for good reasons, with data to back them up.

Some trades are better than others. For instance, trading future maintainability for development expediency now is a losing proposition. You will regret it until the solution is retired. Trading scope (functionality) for cost and/or schedule is also a mistake in the long run, disguised as a short-term gain. Note the pattern: any trade that appears to be a short-term gain but will cause trouble down the road is a trade not worth making

There are two things which you should not trade away for any reason: security and sufficiency. You can trade away completeness at the cost of decreased reusability, but if your solution is insufficient or lacks the necessary security, you have failed.

One of your design goals should be to minimize, as much as you can, the amount of code you have to write. There are many reasons for this, but being lazy or elegant are not among them. The mere existence of code creates costs: it has to be tested, changed, and deployed, repeatedly. If you can get your software components to work harder, you can save these costs. Every hour you spend on design to eliminate the need for additional code is worth well over 10 hours just on the initial version, with continued savings each subsequent version. The less code an application contains, the more reliable it is. There are fewer components that can interact with each other in hidden ways, and it's easier to understand the design.

Primitiveness and elegance go together. More primitive components are easier to write and maintain, and those who follow you in your code will thank you for making the additional effort.

Except for individual methods, primitiveness is a relative measure for which more primitiveness is better (see Section 1.6.6).

To summarize, your design goals are, in no particular order:

- minimize the total amount of code you write,

- reduce coupling, or trade "bad" kinds of coupling for "better" kinds,

- increase cohesion both in the problem domain and in the design, and

- increase the overall level of primitiveness.

You will soon find that 75–80% of the decisions you make are nearly automatic, and will be correct just about that often. Where you will spend most of your time, and most of your late nights, is on the 20–25% of the problem that is hard, new, or unusual. This chapter is written with the idea that you don't need help with the automatic decisions, and spends the vast majority of its time on several hard problems I've encountered on one project or another. I call them hard because, at the time, obvious solutions were not widely known, and I've encountered other developers with the same or similar problems many years later. Along the way, we will use measurement data to make design decisions as we work through the process.

You might not encounter some or any of these hard problems in your application. I don't always see them in mine, either. If that's the case, consider yourself lucky, this time. Next time, you might not be so lucky.

4.1   A COUPLE OF THINGS TO KEEP IN MIND

Design is about solving the problem as best you can. Engineering can help determine if you can do better.

The design patterns and many of the other techniques mentioned in this chapter originated in the world of object-oriented programming. While they are easiest to use in that paradigm, and are always described in object-oriented terms, there are many ways to implement the intent of the pattern without using object-oriented programming. An understanding of intent combined with a little creativity goes a long way.

As you build out your design, you can follow a depth-first path through the application domain, adding application-specific and application-generic abstractions as you go. This approach works best when you work by yourself or want working code quickly. A breadth-first approach is pretty much big design up-front, but works well when multiple teams are involved.

4.1.1   ABSTRACTIONS HAVE A LIFE

Every component in an application has a lifecycle, separate from the lifecycle of the abstraction in the problem domain: it is created, performs its function, and then is destroyed. Creating objects, or allocating space for other types of components, consumes resources. Failing to release those resources creates memory leaks that can lead to crashes. Early on, I took to explicitly assigning responsibility for creation and destruction events for every object in the design. The design analysis technique in Chapter 6 of [5] provides a concise way to determine where lifecycle events, especially object creation and deletion, need to go.

4.1.2   ABSTRACTIONS NEED HANDS AND LEGS

In the problem domain, a *service* (Section 3.2.9) is an externally observable or otherwise required behavior of an abstraction. The work necessary to deliver that service might not be cohesive or may be too complex to implement as a single block of code. Services from the problem domain are broken down into *methods*, a term borrowed from object-oriented programming to describe individual messages that a design component can receive and understand.

Remember, methods are receptors for messages. As with services, there are a few things you need to know about a method before you can write it, or you will have to stop in the middle to figure it out. Since you have to know anyway, you may as well write it down. A good place to put this information is in a comment at the top of the method definition in your code. As you identify individual methods, complete these items.

1. Name the method. This should be a short (remember, you have to type it every time you call it) verb-phrase that gives a clear picture of what the method does. You are naming a message receptor, so use a phrase that senders of the message will recognize.

2. Describe the method in two or three sentences.

3. Identify preconditions, conditions that must exist in the problem domain before the method can execute (be sure to test for these at the top of your method; all sorts of bad things happen if a method fires when it shouldn't).

4. Identify postconditions, conditions you can assume to be true after the method has completed.

5. Identify the invariants, those conditions that do not or cannot change that might influence how this method works.

6. List this states in which this method may be requested.

7. Identify the external data that the method must use to perform its function. These may take the form of required or optional parameters in the method signature, but look out for increased coupling, or coupling that is less desirable. It is better for the method to go get the data it needs, but that isn't always possible. Using a component reference or copy can cut down the number of parameters, but forces the method to know about the structure of the component, which increases coupling between the sender and receiver.

8. Identify the data to be modified and returned to the sender or as a side effect (a value changes, but the sender is not notified of the new value—this often takes the form of an update to a database).

9. Identify the messages sent to other design components.

10. Determine, at a high level, the processing steps for the method. If you already know this down to the level of pseudocode, go ahead write it down.

Focus on making each method as cohesive and primitive as you can. There may be opportunities to turn a set of steps in a method into multiple more primitive methods, especially if that sequence might be used somewhere else (you often discover this much later in the development process, so make the change when you notice the opportunity; it's never too early).

Another good idea is to check the preconditions, the current state, and the user and their role right up front, so you can exit quickly if all of the conditions are not met. This is Michael Nygard's Fail Fast stability pattern [6, pp. 120–122]. The use of mix-ins mentioned in Section 4.3.4 checks to ensure the user is logged in and has the proper role(s) before any function is actually executed. Checking the other preconditions should follow immediately.

The postconditions and invariants make for good test cases for a method. If the postconditions and invariants are true after a method runs, you have met its external requirements. You can use tests of preconditions, postconditions, and invariants to model the behavior of a design before you write the code (see Chapter 6 of [5]).

4.2   SET THE BOUNDARY

Your first task is to set the boundary of your application domain. In practice, this means drawing a closed loop on the problem domain model that separates it into two parts: inside and outside (Figure 4.1, which shows the boundary for the case study). The entities and objects inside the loop are now part of the application domain. Sounds simple enough, but decisions about *where* you draw that loop can have a huge impact on the scope, and therefore the budget and schedule, for your project. Including the wrong entity may dramatically increase the complexity or risk of the project. Excluding the wrong entity carries the same risk.

The best advice is to include those entities that are necessary to deliver on the promises of the project, and nothing else. The more you can safely leave outside the boundary, the better, since everything inside the boundary has to be built, tested, deployed, and maintained over time.

Any entity that is left outside the boundary but has some kind of relationship with one or more entities inside the boundary raises the requirement for some kind of interface.

The next section gives detailed advice on where to place your boundary.

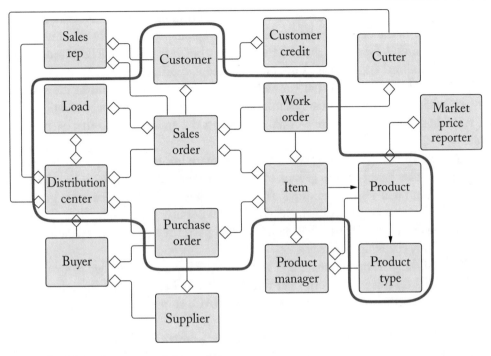

Figure 4.1: Problem domain with boundary.

4.3   BORDER CONTROL

Setting the boundary to your application creates a host of cross-border issues, including communications or relationships that cross it. Mostly, these issues aren't difficult to deal with and generally fall into one of two categories: dealing with people and dealing with other systems or devices.

4.3.1   PEOPLE AS ROLES

One of the things you should have done when defining your problem domain is to group your users into classes. Some authors suggest you create a persona, an avatar for a generic member of each class (an idea stolen from Marketing). Personas may appear as a person or organization in your problem domain model. Sales Rep, Buyer, and Cutter in Figure 4.1 are three examples of what you might see. Each persona is allowed to perform some specific set of operations, and more importantly, prohibited from performing others.

Over the years, many schemes have been proposed that try to balance the necessary security while reducing the burdens on developers and administrators. The industry has settled on Role-Based Access Control (RBAC) as the best compromise [21].

Feel free to use one of the other methods for controlling access to functionality, just be aware that I have found RBAC to be the easiest to implement, especially when using object-oriented programming. Django includes a rudimentary but very effective RBAC tool in the box. It's free, and functional enough that I used it for a recent application.

The idea behind RBAC is that you define a role for every operation in the system. You can assign multiple operations to a single role, or you can create one role per operation. The best advice is to align the roles with the workflows through the application so that each use case is one role.

Users are maintained in the system with just enough information to manage access. Users are assigned to roles based on the groups to which they belong. If a person has the ability to perform more than one set of operations, that is, they belong to more than one group, assign them all of the necessary roles. Allowing multiple roles per user greatly simplifies the process of checking to see if a particular user has permission to perform a particular operation. Easier is better because this checking happens a lot.

When building the operations, I add code to check for the role needed to access it against the current user. I do this in the controller functions (views in Django) using inherited code. I find this to be by far the easiest way to implement RBAC. Your mileage may vary, of course, so use the method you find best.

The point is that all an operation needs to know is which role is allowed access, and then determine whether the current requestor (user, message, service, etc.) has the required role. This works whether your application is all one body of code, or a set of microservices using a separate identity management tool (IAM, for those in the know).

4.3.2   EVENTS ARE JUST MESSAGES

During the "aughts" (2000s), there was a lot of talk about "event processing" and its effects on application architecture. Application architects spent lots of time working event processing components into their designs. For me, it all came to a head when I asked one of these architects what an event looked like inside the application. After an hour, he agreed they take the form of messages. Event processing was a new name for message passing.

In fact, the main job of components on the boundary of the application is to detect events that happen outside the boundary and convert them to messages for whichever components(s) need to deal with them. I realized this when I developed the mathematics to simulate the behavior of a design [5]. Once an event is recognized, and that's all you have, the recognition that an event has occurred, it has to be converted into a message to be handled by a component.

An external event can be triggered when a sensor detects something that it needs to signal, or when a user does something via the user interface. It is convenient, but not always desirable, to lump the user interface into the same category as other types of events because they are generally handled in the same way—converted to messages—but users require far more support than devices, so the user interface is always far more complex.

Once you start to accept that all external events are pretty much alike (except in the required response, of course), the next section makes more sense.

4.3.3   DEVICES ARE LIKE USERS (MOSTLY)

Eternal devices, such as sensors or in one of my applications, tools that process images, behave and can be managed the same way you manage a human user. You want to authorize and authenticate the device, you want to limit what the device can do, you want to record what the device has done. Your user management components already do all that; do they really care that the user on the other end is not a human?

Back in 2010 or so, I was working at a mobile phone company when a project came across my desk that ended up being a major disruption. The business problem was that we were running out of phone numbers and the projected explosion of devices on the Internet would exacerbate that rather quickly. At the time, there were 5 billion devices on the Internet, and projections were for 50 billion by 2030 (I think we've already passed that). The real problem was that we didn't have a good way to uniquely identify every device so that we could authenticate it. We used phone numbers to identify mobile phones, but that scheme wasn't going to work for everything because there just weren't enough numbers.

So rather than solve the stated problem, I suggested we use Session Initiation Protocol (SIP) addresses (which look like email addresses) for identification. The structure of the address lent itself to making use of the domain system being used for the rest of the Internet, and if we let each domain owner name their devices, we'd meet all the requirements for uniqueness. It seemed so simple, until politics got involved. I left before that project got very far, but when I left, I was neck-deep in international standards work pushing a rope uphill. It seems that the Internet of Things (IoT) still has not completely solved this problem, but it's getting closer.

The point of this story is that we use email addresses to identify people, so if we use SIP addresses to identify devices, all of the current authentication mechanisms will work, even those that use public/private key pairs. For purposes of authenticating, controlling access, and recording activity, a device is just like a user.

4.3.4   MIX-IN INGREDIENTS FOR ROLE ENFORCEMENT

An easy way to implement a role check is to use a function decorator (supported in Python, for example) or inherit from a class that performs the check (a "mix-in" class). I use the mix-in class approach so my Django views (the controller in the MVC model) inherit from the appropriate mix-in (see Figure 4.2). There is one mix-in for each role. This isn't as bad as it sounds—my current application has 4 or 5 mix-ins, each with 5–10 lines of code. Each mix-in inherits from a class that authenticates the user.

These mix-in classes return True if the current user has the role required for the function, and False if not. If the user has not yet been authenticated, the base class directs them to the

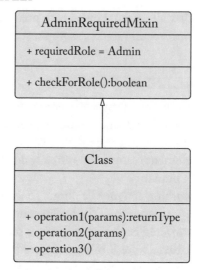

Figure 4.2: Checking for role using mix-in class.

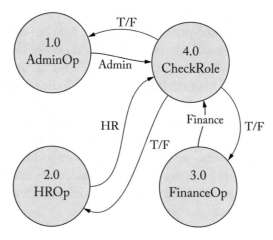

Figure 4.3: Checking for role using function.

login page. Adding a new role is a matter of creating the new mix-in and adding that role to the database, a process that typically takes about five minutes.

I prefer to use classes and inheritance, but this could just as easily be handled by calling a function with the required role as a parameter and getting back a True or False (see Figure 4.3). The advantage to this method is that only one such function is required for the entire application.

4.3.5   PASSWORDS?

Like everyone else, I built applications that stored passwords, until a few years ago when I converted a current application to use a third party identity provider. That sounds fancier than it is; we identify users by a Gmail-managed email address and validate that email using Google APIs. There are many other options available for third party identification, including Facebook, LinkedIn, and others.

Storing passwords is a security risk, even if they are encrypted. In the 1980s, convenience stores discovered that if they didn't keep large quantities of cash in the store, even in a safe, their risk of being robbed dropped considerably. The Target breach, and others, taught the rest of us that if we didn't keep payment card information in our systems, the inevitable data breach is rendered relatively harmless. So it is with passwords. Third-party identity providers maintain passwords on your behalf, but you have to be able to trust them. Most social media platforms, including Google, Facebook, and LinkedIn, provide third-party identification services, usually free. We chose Google because people already have a Google ID or can easily get one for little or no cost (not all cost is monetary, many people just don't want to be on Facebook or LinkedIn).

Passwords may soon be a thing of the past. Recent advances in identity verification have led to the notion of a self-sovereign identities (SSI) using public/private key pairs. Use of SSI requires a supporting infrastructure, which at the time of this writing, was mostly built on top of a Blockchain or equivalent distributed ledger. Also at the time of this writing, a ledger-less method called Key Event Receipt Infrastructure (KERI) [29] was being developed that looks very promising.

Either form of SSI will eliminate the need to maintain user authentication code as you can rely on publicly available utilities to validate that someone claiming to be one of your users is actually one of your users. All you will need to do is maintain the set of recognized users, including one of their public keys, and the set of roles to which they have been assigned. The resulting security infrastructure will be far stronger than anything using passwords.

4.4   ACKNOWLEDGE THE CRUD

The mere existence of entities inside boundary of the application domain creates the need for four maintenance functions: create, read, update, and delete, or CRUD. These are especially important in data-strong situations since the primary purpose of the system is to model information about these entities, then use that information to do its job. Even function-strong and control-strong situations need to provide CRUD functionality for entities in the data model.

To be honest, this is only mostly true. While there is a need to be able to create, read, and update data right out of the gate, I have never included delete functionality in the first release of any application, for a couple of reasons. First, deleting something is the one function that can wait. Even if you want to delete something and can't, the rest of the system will function as it should while you safely ignore it. Second, and more important, it takes a certain amount of experience using the application to figure out what "delete" really means.

There are two kinds of delete: physical delete where the data is removed from the data store, and logical delete where the data remains but is marked to be ignored by the system. Physical deletion gets complicated very quickly when you start navigating the relationships between entities. Many relationships imply foreign key constraints such that, when you go to delete a customer, for example, you have to also delete every transaction that refers to that customer, and now you see the problem.

You never, ever want to delete a transaction that was legitimately completed because doing so would leave a hole in the auditability of the business records. There are legal and other requirements for records retention, and business transactions are at the top of every list. So, no physically deleting customers. The same applies to most other entities.

To solve this, developers created the concept of logical deletion, in which data is marked in such a way that makes it easy to simply ignore. Logical deletion has the wonderful property of not affecting in any way the existence of any other entities that have a relationship with the deleted data, unless you want it to. Logical deletion is so prevalent it is a full-fledged design pattern.

Depending on the size of your data model, the amount of CRUD functionality you need to implement can be daunting. Fortunately, it is almost free when you build using a framework such as Django or Rails. This nearly free functionality is one of the main reasons I make all of my applications web-based; the ability to use them from anywhere is another.

4.5   COMBATING VOLATILITY

There are two sources of volatility: the problem domain and other components. In this section, we look at each of these and explore ways to mitigate their effects. You can't eliminate it. Section 1.6.8 defines a measure of volatility to help make these decisions.

It takes experience, both in software design and in the problem domain, to accurately guess where volatility will originate. Maybe you felt that the client community just didn't have a good handle on the workflows or the haven't settled on navigation, indicating that the user interface could fluctuate, at least in the near term. The nature of the problem provides clues about where volatility likes to hide. In data-strong applications, the user interface and business logic will be the most volatile; data models are surprisingly stable. In function-strong or scientific software, the mathematics and algorithms in the filters will change more often than anything else. Even if only the data objects change, the filters will have to change. In control-strong or real-time software, every new device and every new version of old devices creates volatility.

4.5.1   ISOLATION

The main strategy, really the only strategy, for combating the effects of problem domain volatility is isolation. You design your software so that the parts of the system that vary the most are isolated from other parts of the system so as to lessen the impact of a change. You will never completely isolate volatility, but you can isolate it to much larger extent than you might realize.

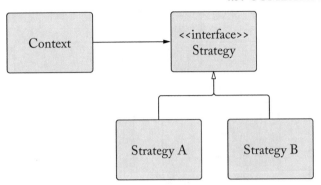

Figure 4.4: Strategy pattern [9].

The Layers and the Model-View-Controller patterns are designed primarily to isolate parts of the application from changes in other parts. The main benefit of the Microkernel pattern is the way it isolates any part of the system from every other part.

The Gang of Four patterns catalog [9] contains many design patterns you can use to isolate changes. For example, the Strategy pattern (Figure 4.4) delegates an operation to a set of abstractions, each of which implements a version of it. The abstractions are related to a generalization that is known to the owner of the operation. This allows the choice of version to be made at runtime.

Another way to isolate volatility is to create new walled compounds (layers). You can take a group of abstractions that might be volatile and likely change together and lock them behind an interface using the Façade (Figure 4.5) or Bridge pattern. The rest of your application sees only the very stable façade.

Volatility can also result from exposure to other abstractions that may themselves be volatile. Volatility follows lines of coupling, and its degree and effects are related to the type of coupling. This suggests that reducing the less desirable forms of coupling might reduce the exposure to volatility. Decoupling abstractions is just another form of isolation and uses these same tactics and patterns.

4.5.2 DYNAMIC ATTRIBUTES

In one of my recent applications, two of the models were subject to high levels of volatility in their structure. Some of the more interesting attributes about a brain tumor are the various genes and their expressions tracked by researchers. Not only do different types of tumors have very different structures, it seems that every conference or paper introduces a new gene. A new gene is a new attribute, so I was making table changes on a regular basis. Any change to a database table requires changes to views and controllers that are always more painful than other kinds of changes. Almost any effort to reduce the effects of these kinds of changes is worth it.

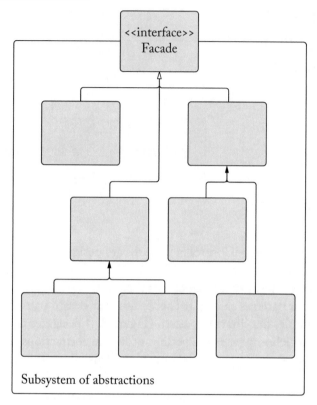

Figure 4.5: Façade pattern [9].

To combat this form of volatility, you have three options.

1. Put all of the attributes from all types of the entity in the same table, making rows that are sparsely populated.

2. Create separate tables for each type (even subtypes or specializations need their own tables).

3. Use some kind of dynamic attribute scheme.

I should note that only the last option also solves the problem of frequently added attributes.

I implemented dynamic attributes (they're not "user-defined" because we don't let users anywhere near them) so that adding a new gene requires nothing more than setting up a new attribute and configuring the code that drives the user interface. The new attributes take effect when the user refreshes their browser. Implementing dynamic attributes was a huge task, but it has proven to be a net time saver in the first year of its existence.

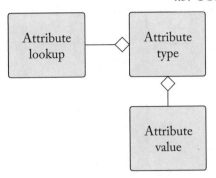

Figure 4.6: Standard EAV pattern.

Fortunately, there are a couple of patterns that help address this kind of problem. The main pattern is Entity-Attribute-Value (EAV) and serves as the basis for the others (Figure 4.6). The basic idea is to build a data structure that stores entity-attribute-value triples for a set of defined attributes. Adding a new attribute is simply a matter of creating a new attribute definition and populating the entity and values. The pattern was first used to handle the large quantities of unstructured data in medical records [30].

Like most things software, this is easier said than done, and the basic EAV pattern works only when all of the attributes are attached to different instances of the same entity. We used this pattern, with an important modification, in the case study.

That brain tumor application had a couple of additional complications. We needed to model multiple kinds of tumors as well as some conditions that were not tumors. In addition, we needed to model random types of outcomes, each with its own set of attributes.

Another issue is that different attributes have different types: integer, floating point, string, lookup or reference, Boolean, and date or date/time. This creates a design problem for the table(s) that actually store the values, leading to another pair of options. I also had to provide enough information to allow the user interface to configure itself on the assumption that most of the time, most attributes would not be used (the Attribute Use table).

I solved all of these problems with the classes shown in Figure 4.7. I chose to group all of the values into tables by data type, so all this integer values are in one table, for example. There are alternative solutions, such as using columns for each type or converting everything to and from a string on write or read, but I'd used this particular solution before. Lookup or reference values are stored in a table (Attribute Lookup) of possible values grouped by attribute. Each row contains the order in which the values should appear in a selector, the default option, and labels and actual values for all of the option.

The code to manage all of this is assigned to the Attribute class and handles saving of values, audit trails, and tables where the changes need to be reviewed by someone. This class is a major part of the application-generic domain in my latest application.

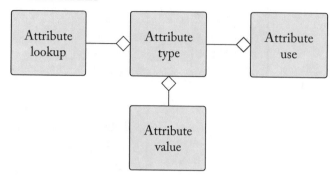

Figure 4.7: Alternate EAV implementation to support multiple uses.

Should you need user-defined attributes, find the pattern that most closely matches your particular situation and go from there.

4.6   IT'S OK TO INHERIT FOR REUSE

Despite what some object-oriented purists say, creating a gen-spec relationship to share some code is a perfectly valid thing to do. Remember, the idea is to write as little code as possible, and sharing code is a very effective way to reduce the required code.

This applies when not using object-oriented techniques, too. In fact, it's a bit easier because factoring out common code means putting it in a method of its own. You still create a new dependency, so the design measures are the same or very similar in both cases.

Figure 4.8 shows the structure of one of the model classes, Exam, in one of my larger Django applications (actually a collection of five applications that use the same set of models).

The tricky part is to decide what you can factor out and reuse. When comparing to design components, measure the similarity between them. Even if they aren't similar when you look at the whole component, look at various parts, and if those parts are similar enough, consider splitting them out into a common generalization component. Each component is now dependent upon the generalization, but they are not dependent on each other (unless that was already the case).

In the example, the set of attributes are always the same in both tables. The implementation for the audit table classes are actually very simple and contain only a foreign key reference to the instance of the main class. Figure 4.9 shows the structure of the Exam and Exam Audit tables with the attributes factored out. I use this particular pattern a lot in that application as it saves a lot of time and ensures that the set of attributes for a class and its audit table are always aligned.

The structural similarity between the Exam and Exam Audit Table is high, since the attributes make up the vast majority of the properties, even though the actual measure value is

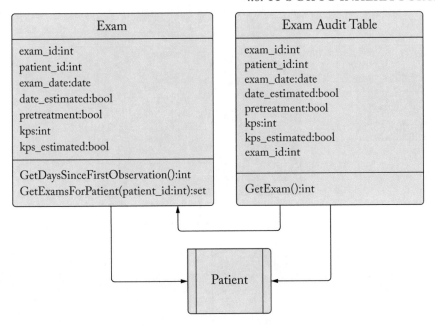

Figure 4.8: Original exam structure.

fairly low. Recall from Chapter 1 the formula for similarity:

$$sim\,(a,\,b) = |a \cap b| - |a \backslash b| - |b \backslash a|.$$

The similarity calculation is:

$$sim\,(Ex, ExAT) = 7 - 2 - 2 = 3.$$

The original complexity calculation includes two relationships to the Patient class and one from the Exam Audit table to the Exam class. Not shown on the diagram is a gen-spec relationship from each class to a LogicallyDeletedTable class to implement logical deletion of exams, as well as a gen-spec relationship from each class into the Django model structure.

The similarity calculation after the factoring is negative, indicating no similarity at all, which we expect. The two relationships to the Patient class have been replaced by one, as have the gen-spec relationships to LogiciallyDeletedTable and the Django model structure, but we've added a new gen-spec relationships, so overall complexity is actually reduced slightly. Between the reduced complexity and the reduced maintenance effort, the design change appears to be a good decision.

This same design pattern was used over two dozen times in the application. You can see how small changes can add up.

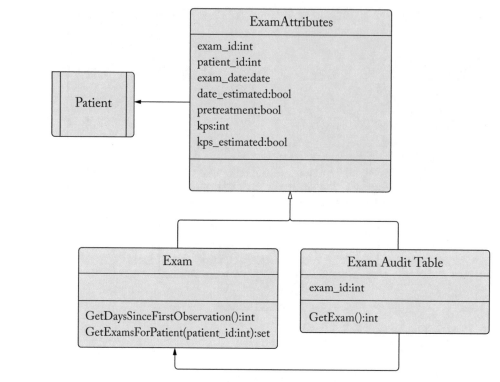

Figure 4.9: Exam tables with attributes factored out.

4.7   A STRATEGY FOR SUBSTITUTION

If you have two design components that have different structures but can sometimes stand in for one another, make a common generalization and make sure the code expects an instance of the generalization. If you follow Liskov's Substitution principle in your code, you can safely substitute a component for its generalization at runtime. This is an application of the Bridge pattern from [9], but not one the pattern was written to cover. With experience, you will recognize situations where you can use a known pattern, even if the pattern doesn't include your specific situation.

Generally, this requires that both original design components support, at a minimum, the set of methods required by the consumer code. You can easily enforce this by making the common generalization an abstract class or interface in which only the required methods and their signatures are defined and leave the implementations to your individual design components.

Similarity can be used to figure out how much the two components have in common. Unlike the previous section, however, where you moved actual code into the generalization, all you're going to move here is structure in the form of method names and signatures. The code that

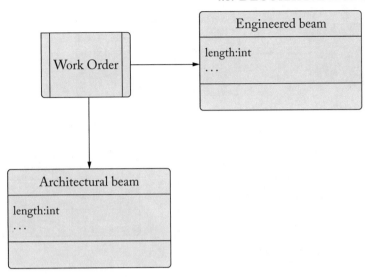

Figure 4.10: Original work order design.

accepts the generalization will automatically redirect to the version of the method on whichever design component you use.

I have done this in several applications over the years. In the case study, different kinds of wood products, with different structures, were used on work orders to create custom lengths to support an order (Figure 4.10). The work order needed only a simple data structure and one or two operations. The required data was common to all wood products, and we had to make sure that all of the wood products supported those operations (Figure 4.11). The redesign made the work order code much less complicated.

4.8 DECORATIVE MODIFICATIONS

There are times when you want part of a design component to behave slightly differently. There are two ways you can do this: create a new function for each set of behaviors (which almost requires a form of method overloading), or use a Decorator or State pattern [9]. A decorator (Figure 4.12) adds new functionality to an object of a class without modifying the class itself (this object-oriented technique is a bit more difficult to replicate in other programming styles, but it can be done).

To create a decorator, create an abstract class or structure that defines the name and signature of the method who's behavior will change. Design the function that uses this behavior to accept the decorator. Split out the function from your original design component into subtypes of the decorator, one for each set of behaviors. At runtime, substitute the subtype with the desired behavior in place of the decorator.

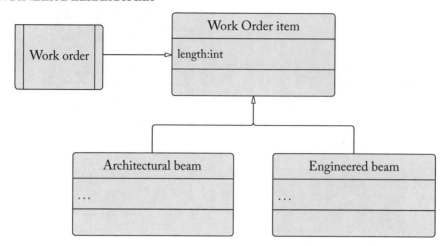

Figure 4.11: Work order using bridge pattern.

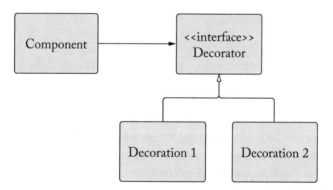

Figure 4.12: Decorator pattern [9].

Some languages, such as Python, support custom decorators for functions. With some work, you can use this technique instead of the object-oriented method.

When dealing with medical images, the DICOM standard is not really standard. There are four or five MRI scanner manufacturers, and each makes proprietary changes to the standard that can seriously mess up your ability to read and process the resulting images. Among those who deal with medical images on a regular basis, it is well known that nearly all software that processes DICOM images have separate paths through the logic based on the manufacturer of the source scanner. Even the best tools fail if the manufacturer is unknown. Figure 4.13 shows the Decorator pattern being used in an image processing scenario.

The State pattern [9] can also be used to alter the behavior of a component at run time (Figure 4.14). The State pattern requires that you create an abstraction that represents various states of the original component. This abstract state is a placeholder for other abstractions that

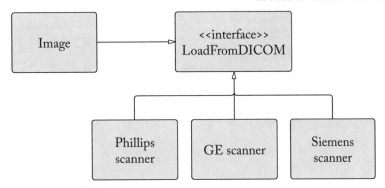

Figure 4.13: Medical image processing using decorator.

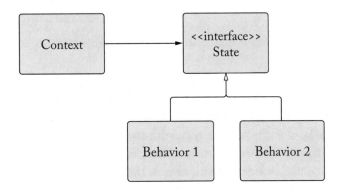

Figure 4.14: State pattern [9].

implement particular states. Note that not all states need to be dealt with, only those in which the behavior changes.

Your choice of which to use depends on circumstances. If you want the ability to choose an algorithm or set of logic based on external information, use the Decorator pattern. If you want an object to behave differently in a state, but to always behave that way when in that state, use the State pattern. I use the Decorator pattern quite a bit but rarely use the State pattern because the situation doesn't come up very often.

4.9 ADDRESSES EVERYWHERE

A typical sales order in a business selling to other businesses (could be wholesale, might be retail) has the potential for up to three different parties to play the role of a customer: there is the sold-to customer that authorizes the sale, the bill-to customer that will pay for it, and the ship-to customer that will receive the goods. Each customer will have an address.

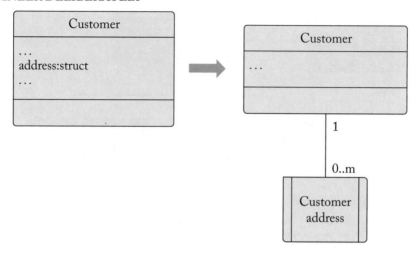

Figure 4.15: Add aggregation when there can be more than one instance of a component.

There are two degenerate end cases, and a whole lot of in-between. On one end, all three roles are filled by the same entity and all three addresses will match. On the other end, each customer is a separate legal entity with its own address. More typically, the same customer fills all three roles, but at least one of the addresses differs from the others.

It is common practice, based on the number of times I've seen it done, to make the address either a single attribute, or a set of attributes, of an entity such as Customer. This implies a 1:1 relationship (an aggregation) between a Customer and an address. The reality is that a Customer can have some random number of addresses, making the relationship 1:m.

In data modeling, there are only three important numbers when it comes to relationship cardinality: zero, one, and more than one. Zero is easy and we'll ignore it. When the cardinality is one, and is guaranteed to stay one, you can implement the relationship by embedding the attributes of one entity into the attributes of the other (Figure 4.15 left). If the number is more than one, or even if there's a chance that it might be later on, you have to implement each entity separately and create an instance connection from the M-side (the Address in this case) to the 1-side (the customer in this case) (Figure 4.15 right).

When we first designed the case study application, many of us encountered this problem for the first time. As we dealt with it, we came to calling it the "address problem" even when we encountered the situation with two entities that were not addresses or customers. A better name might be the "multiple component" problem.

The pattern here, when an apparent 1:1 aggregation is or could be 1:m, can be hard to spot. I've had clients swear the business rules only allow for one whatever at a time even when there are many cases of more than one whatever in the real world. Be suspicious of 1:1 aggregations, they are a minority in the world of data modeling.

Another example is the association of a medical image with a biopsy or resection in my research lab application. Originally, the users assumed there would be only one image per surgery for recording the location of the biopsy or resection. After some experience, they needed multiple images, from multiple sources, to track the location. Clearly, what was a contains situation became an association. The change required removing several attributes from the treatment table and creating a new table to handle the m:n relationship between an image and a surgery event. This new table had the additional attributes to store the coordinates in two coordinate systems: voxel location (pixel, slice), and real-world relative to the patient position.

You can take the safe route and always implement any aggregation as if it were 1:m. Implementing a 1:1 relationship in this way is not harmful, but it will require more work, both now and down the road. Implementing an aggregation as a set of attributes as if it were a 1:1 only to find that it is really a 1:m requires a lot of work to fix, work that has to be done right now. So, choose wisely.

4.10 DEALING WITH THOSE PESKY M:N RELATIONSHIPS

Many-to-many relationships, denoted m:n, are a problem for software developers. There is no direct way to implement them that also provides an easy way to navigate them, that is, find all of the components given an aggregate, or to find the aggregate given a particular component. Even in databases, there is no direct way to implement an m:n relationship without adding a table.

4.10.1 THE NORMAL SOLUTION

Keeping track of the participants in an m:n relationship requires maintaining the current set of participants on both ends, and keeping both ends in synch. It is far easier to add an abstraction to represent the association and model it as a pair of 1:m relationships. This method of implementing m:n relationships is so common it has become a pattern, Association Table Mapping [13, pp. 248–261].

As shown in Figure 4.16, the m:n association is replaced with an abstraction that represents the association but has an m:1 relationship with each of the original abstractions. In the data model, we implement these as associative tables to handle cases where the association itself has attributes or behaviors (but all m:n relationships require these associative tables in the physical database).

If you prefer to decouple each of your original abstractions from the association, you can design the associative abstraction so that it can find either side of the relationship, but have the related abstractions be ignorant they're even in a relationship. The equivalent graph structure design (Figure 4.17) has the edges contain all of the references, with the nodes knowing nothing about edges. Navigation of this sort of structure requires the ability to quickly search the set of

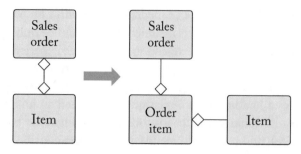

Figure 4.16: Change m:n into a pair of 1:n relationships.

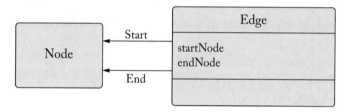

Figure 4.17: Decoupled directed graph structure.

associations for a node abstraction, but that might be worth the extra effort if it lowers the coupling of the entire design or isolates some volatility.

We used a similar structure for truck load planning in the case study. In that application, the edges were the main thing of interest and the nodes were just locations. By using this structure, we could completely decouple the addresses from the process. The load plan consisted of a string of addresses linked by edges that contained distance, cost, weight, and sequencing information.

If the association itself has significant behavior, or has the potential to be a useful abstraction in its own right, you should consider implementing a structure similar to the shipment described in the next subsection.

4.10.2 WHEN NORMAL DOESN'T WORK: THE LOWLY SHIPMENT

Conventional wisdom, and nearly all of the literature, about the structure of a sales or purchase order transaction assumes that the order is a relationship between two parties and some number of items. As is often the case, conventional wisdom is wrong.

We originally designed both sales order and purchase order for the case study application following conventional wisdom (Figure 4.16). Then reality started getting in the way. First, it was how to handle backorders where some items are shipped on a later date. Some orders were shipped directly from the supplier to the customer or items on a single order could go to multiple locations. Some deals were structured rather strangely, unless you were familiar with the lumber business. Finally, we had to support logistics and load planning. It quickly became clear that the conventional structure for an order just would not work.

An item is backordered when sufficient stock does not exist at the time the order is shipped to fill the order. This can mean that part of the order is filled now and part later, or that all of some item is shipped later. If you've ever ordered from Amazon, you've encountered this situation. The basics for the order, the shipper, receiver, and all of their document numbers, do not change, only the quantities and ship dates change. After a backorder or two, you can have more than one ship date for a single order, each of which applies to a unique mix of items and quantities.

A direct order is a case where a sales person sells a set of items to a customer but has the supplier ship them directly. The most common way to handle this is to create both a sales order and a purchase order, each with its own set of items. This is duplication, and belies the fact that there really is only one transaction. Keeping the two orders synchronized requires a lot of work on the part of the software.

I've already described an order structure unique to the lumber business, the "to be determined" or TBD order. In a TBD order, a large quantity of lumber is ordered at one time, to be delivered over a period of time, but the actual mix of items, the actual quantity, and the shipping dates are not set at the time the order is placed. When a shipment is ready, the seller will send an advance shipping notice (ASN) to the buyer with the details of the next shipment. The buyer then works to sell the goods before they reach the warehouse by splitting off parts of the shipment to customers along the shipping route, so even a single TBD shipment could result in multiple sales orders, all of which are tied together.

In another type of order, not that uncommon it seems, a customer places an order for items to be shipped to several locations. This was most often done by chains of small hardware stores to take advantage of bulk pricing.

The logistics function planned truckloads and delivery routes for items shipped to customers. Some of these routes covered multiple cities, one or two covered multiple states. Shipping lumber is expensive, so loads had to be planned to avoid unnecessary travel, especially for an empty truck.

After many, many arguments, we came up with one solution that solved all of these problems. We split the items on an order into a separate entity we called a shipment (Figure 4.18). A shipment could relate to one sales order and/or one purchase order. The orders maintained the relationships between the parties and the basic terms of the transaction. The shipment carried the ship date, source and destination shipping addresses, and items, including cost and price, each in one or two currencies (a direct shipment from a U.S.-based supplier sold by a Canadian distribution center to a French customer gives you three currencies on a single transaction— there is an island off New Brunswick that is part of France, and still uses the French Franc, or did at the time, they may use the Euro now).

To handle a backorder, the backordered items were simply moved to a new shipment attached to the original order. TBD orders were dealt with at the shipment level, and when

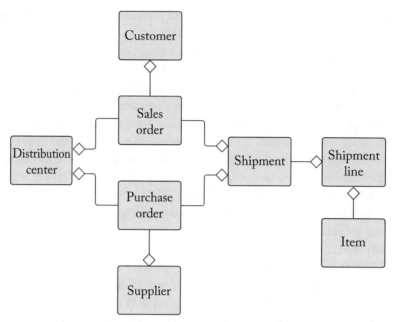

Figure 4.18: More complex m:n structure using intermediary entity (Shipment).

some of the shipment was sold in route, a new shipment was created and attached to both the new sales order and the original purchase order. Direct orders were simply shipments with both a sales order and a purchase order. Logistics used shipments for planning, and we eventually defined a shipment as "a collection of items going from point A to point B at the same time on the same truck." The load planners could split a shipment into two separate shipments when they needed to split the load between two trucks, but rarely did.

This model worked out so well it stood up to the arguments presented by every new member of the team. When a new member objected, I wouldn't disagree, I would simply have them present their better idea. They always talked themselves into supporting the shipment design; it became rather entertaining for those of us who'd been on the project for a time.

I don't think I will ever implement an order that doesn't include a shipment as a separate entity. Once you get used to a solution, there is no going back, and so far, nothing better has come along. When we built that application, Amazon didn't allow splitting orders, so we were among the first to use the concept. Now, just about everyone does.

Several years later, I discovered this very pattern documented in [31, pp. 151–196]. We implemented a modification that skipped the Order Item object. In our world, an Order contained one or more Shipments, each of which contained one or more Line Items. I don't see the Order Item entity in the pattern as necessary or useful.

4.11   THE TRANSACTION AS TIME CAPSULE

The address issue raises another interesting challenge as they tend to change over time and you need to know where those goods were shipped last June, or three years ago last June. But it's bigger than just addresses.

Business transactions, such as a sales order, reflect the state of the business at the time the transaction occurred. Some of that state needs to be preserved after the transaction has been completed. The addresses of the various parties involved are one example of such state information. Other examples include the terms of the order, the prices charged, the items ordered and shipped including their cost, the dates of various events, the units of measure used, and the currency exchange rates in effect at the time. Don't forget values found in lookup or reference tables as those, too, will change over time.

The challenge is to figure out how much of the business state needs to be captured for a transaction. The solution is pretty easy: stamp the transaction with values that indicate the state at the time. This can lead to large transaction rows, sometimes very large, but it's a trade usually worth making. While this may look like duplication, it allows the transaction values to remain static as the original values (the reference data) change. This is the main reason to never physically delete anything that might appear on a transaction.

4.12   BEWARE OF SHARED RESOURCES

During a second job interview at Amazon, the interviewer let on that their biggest problem was keeping millions of frontend servers updated with fluid data such as the balance on hand of an item in inventory. The problem was particularly acute when the balance on hand was running low. I was somewhat surprised by this revelation because we'd solved that very problem more than 10 years before.

The case study includes multiple database instances working in a distributed network, but keeping a central database apprised of all transactions through replication (this is a difficult problem in its own right, as you will see in the next chapter).

In our case, the issue was the customer credit available when a sales order was released to the warehouse for fulfillment. For a given customer, each distribution center sold to one or more local locations, each of which was a separate ship-to customer. If the ship-to customer was part of a national chain, there was only one corresponding bill-to customer, with one credit account, shared by all locations.

To see the problem, let's do a little thought experiment. Say we have two distribution centers that sell to local locations of the same bill-to customer. These two locations share the credit line for the bill-to customer. Let's say, for purposes of illustration, that the credit limit is $100,000.

Distribution Center A releases an order for $20,000 after checking its local copy of the credit balance to ensure there is more than $20,000 available. The process of releasing the order deducts $20,000 from the local copy, leaving $80,000, assuming it started at $100,000, then dutifully replicates the new value to the central database. Distribution Center B does the same thing at about the same time. Both local copies of the credit balance show $80,000 remaining as does the central database. All of these are wrong, the correct value is $60,000. Further, there is no design that can recover from this kind of error.

Like inventory balances, this isn't a problem if there is a large credit limit to start with, but it gets to be a real problem as the available credit approaches zero.

Our solution was to assign updating of the available credit balance to the central server, which then pushed out the modified balance with each update. Local databases operated on the assumption that the local copy of the credit balance could be wrong for a short time. As a sales order was released at a distribution center, a credit-consumed message was sent to the central database that was processed in the order received, ensuring that orders released before the limit was hit were approved.

There were multiple ways to do this, and we chose the easy way out using database replication and a trigger on the central database. As a sales order was released, a row was written to a credit-consumed table that was replicated to the central database. There, a trigger would fire for each new row to update the customer credit balance, which replication would then push out to all of the distributed databases.

I would not do it that way again. In fact, because of my experience with that project, I don't allow database triggers, nor do I allow stored procedures that do anything other than return a set of rows. Relying on the database for this sort of business logic raises a number of issues.

- It violates the rules of the layers architecture that service layers don't know what happens, or is supposed to happen, in application logic layers.

- Database triggers and stored procedures are notoriously difficult to subject to configuration management. You can manage a text file that creates the trigger or stored procedure, but you can't prevent the trigger or stored procedure from being modified in production.

- When a trigger is supposed to appear in only one of several databases in a network, having it show up in a local instance can create bugs that can take days to track down, and even longer to fix.

Today, I would implement the credit consumption message directly as an asynchronous message between the distribution center and the abstraction assigned the responsibility to maintain the credit limit. This could easily be implement using a message bus or some other third-party tool.

There are two lessons from this section.

1. When you have multiple users of a shared resource, whatever that might be, assign one instance to write and let the others be read-only. This is the One-Writer, Many-Readers pattern, but I haven't seen it in any patterns book.

2. Keep business logic out of the database, and don't listen to your database administrators. I once built an entire application using database triggers, stored procedures, and built-in forms, so I know it is possible, but it is not something I will ever do again.

You might be thinking that you won't encounter such issues in function-strong and control-strong problems. For the most part, you'd be right. However, control-strong problems come with real-world timing constraints and constant event notifications that create very similar kinds of issues. Function-strong problems present opportunities for parallel processing that create situations very similar to the one in this section. Shared resources, especially scarce resources, present special design problems in any kind of software.

4.12.1   LOGS AND AUDIT TABLES: NOT JUST FOR AUDITING ANYMORE

If you've ever encountered a problem in production and wondered how it might have happened, you know the value of a good set of logs. There has long been debate about just how much information you need to store as a record of past actions. The answer is simple: just enough to be able to troubleshoot most issues that come up, but no more. This is easier said than done, of course.

When building data-strong applications, I find that I often need to be able to explain how the state of an entity came to be the way it is. Seeing the history of the values of that entity's attributes provides a lot of information about how the software might be misbehaving, especially if that history includes the method that made the change. Think of this history as permanent print statements that can also capture the time a change was made and who made it.

There are lots of ways to implement logs. I've taken to using dynamic code built into the language I'm using to build generic log code that can be inherited by my models. I've used this technique on a number of applications, and find that I have to write the logging code only once per language. Over the last 20 years, I've written it in C++, C#, and now Python (which would be easy to replicate in Ruby). The logs created using this technique take the form of audit tables that parallel the tables for the models (Figure 4.19). This all works by overriding the Save() function on the model. If all you need are text-based logs, there are any number of libraries in many languages that make them easy to implement.

In my research application, we took audit tables one step further. One of the requirements was that a second person review changes to data for most of the entities in the system. I implemented this by enhancing the basic audit table code to intercept a change and write it only to the audit table marked as pending. The application provides a set of functions that allow authorized users to view and accept or reject changes. Authorized changes are then made to the main model table and the audit table row is marked as approved. The model code has no idea whether there is an audit table or that changes require a second review; it's all built into inherited code that is part of the application-generic design domain.

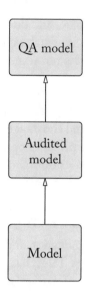

Figure 4.19: Audit and QA table structure.

4.12.2   DATA ACCESS

Since the beginning of software development, the structure of data in permanent storage has differed from the structure of data as we use it in software. They were the same when COBOL programmers used punched cards to store data and close when they used the IMS database on the mainframe, but they've been growing further apart since. First, there was the advent of the relational data model that is so prevalent today, then came object-oriented software and the need for object-relational management (ORM) tools.

The use of frameworks such as Django, Rails, or Hibernate hide this constant reformatting of data deep in layers of abstraction. This makes it very easy for programmers to navigate data stores at the cost of time and space (memory). Personally, I find this tradeoff well worth it and use the Django ORM framework in my scripts as well as my web applications.

Regardless of whether you create it yourself (and I have, multiple times) or use a tool, or whether you use a relational data store or some NOSQL data store, the bottom layer of your architecture is where this function lives. In some cases, this functionality is the bottom of the bottom layer as you may have data integration tools that create virtual data stores, meaning you have no idea where the data is actually stored.

One of the first things application developers learned about architecture was to separate the data access from the rest of the application as rigidly as possible. Doing so allowed not only changing the database management system for another, but different objects could be sourced from different types of data stores and the application code was none the wiser.

Today, few developers pay much attention to where data is actually stored. Enterprise information managers go to great lengths to make this possible. Even if you don't have teams of people to do that for you, the lesson is still valuable: keep your data separate from your logic.

4.12.3   DELEGATE, DELEGATE, DELEGATE

The term "delegate" has a lot of meanings, but the one we use here means relying on code you don't have to build as much as you can. I currently program mostly in Python using the Anaconda build, mainly because it comes prepackaged with most of the third-party libraries I need for scientific computing and image processing. It also provides a handy way to manage the configuration of my programming environment. Ruby programmers are familiar with tools that do the same thing for Ruby and Rails. NetBeans and Eclipse sort of help do that for other languages, but they're integrated development environments (IDEs) and I prefer a text editor and the command line (long story).

Once you enter the world of Open Source software, there is no going back. Open Source programmers solve a problem, then contribute the code to the broader community. Users of the code expect it to work as advertised and are ruthless in rooting out bugs. If one package doesn't work or is a bad fit, there are very likely several other packages that do work and fit better.

On one project, I needed a way to check for cycles in a directed graph. This is a fairly simple thing to do and there are many algorithms available, so I could have written it myself (and have in the past). Before I started down that path, I took a quick look through the Ruby Gems (I was using Ruby) and found one based on a Python library that worked like a charm. It took a few minutes to install and learn how to use, but the problem was solved.

This idea works equally well in nearly every other language. In fact, one of the reasons to pass on using a language is the lack of third-party libraries. In my JavaScript code, for instance, I make use of several libraries and code frameworks, including JQuery, D3 (excellent graphics capabilities), NumJS (NumPy for JavaScript, a really nice numeric processing library), and a few others. It even works for HTML; I rely on third-party style sheets to provide the basic look and feel of the web pages.

As with any third-party code, DO NOT MODIFY IT! Sorry to yell at you, but you need to hear it. Override the functionality if you need to change it, but keep your hands off the original code. Once you've changed it, you can no longer update it and you're stuck.

My point here is that as long as the code works, it's easier if someone else writes it. You treat the features of the library as a black box and you shouldn't care what goes on inside, unless it doesn't work quite right. Your job is to get an application up and running; you don't have the time, nor likely the expertise, to dig into library code.

4.13   SIMPLIFY THE MODEL

If you've done the work right the first time, there won't be much you can remove during a second pass. However, designing and writing code is a lot like writing a book (more similar than you can imagine) and you will always find a tweak or two during a second pass. Your main goal is to reduce the complexity you added during the design as much as you can, but no more. Solve the problem correctly first, then make it fast, then make it elegant, in that order. You typically won't have time do all three, and elegance is the safest to sacrifice.

As you work through the design, you will encounter something that doesn't feel quite right, or wonder what you were thinking. When you do, look at alternative designs, take the measures that help address your design goals, and pick the best option. If you're lucky, you did that the first time. Often, though, you will need to make a change. Hopefully, if you did the rest of the system carefully, your change won't involve too much of the design.

4.14   TEST YOUR DESIGN

There are a number of ways to test a design before you have code to run. Rebecca Wirfs-Brock [32] presented the idea of role-playing using class responsibility (CRC) cards. Peer reviews are also proven effective at finding errors in a design. Chapter 6 of [5] introduced what amounts to an algebra to describe the structure and analyze the behavior of a design.

I created this technique during a project in which I had eight weeks to code and test a design that would take a minimum of six weeks to get to a state where I could test the code. There were a number of risks associated with the project, the time crunch being one of them. The project's purpose was to optimize resource allocation for construction jobs over a three-year planning horizon. Resources were expressed in terms of hours required for a given construction craft, such as plumbing, electrical, or millwright. There was always more work than resources, so the goal of project was to optimize the amount of work and the resources used as much as possible.

One of the constraints of this project was that it had to run in the MS-DOS side of a personal computer that also ran Windows. In those days, Microsoft Windows was not a full operating system and ran on top of MS-DOS. This meant that the program and all of the data had to fit in less than 360 KB of memory. (Remember those days?)

The best data structure to solve the problem joined a craft, a job, and a day, and contained the hours required for that combination. The algorithm basically moved these nodes around the schedule using the optimization strategy chosen by the user (minimize overtime, use a fixed percentage of overtime, or determine how much of each craft was required to get all the jobs done). While small, there were over 19,000 nodes to deal with, plus the code, so space was a real concern.

I knew I didn't have time to start over if the idea failed, so I came up with a way to compute the required memory footprint and simulate a run using math to make sure the algorithm was up to the task. A simulation is set up by specifying the behaviors of each object in terms of pre- and post-conditions for each of its methods (you should already know those—see Section 4.1.2) and which objects need to be in scope when the simulation occurs. The math then models the message passing between objects, determining when a message will fail. The simulation ends when the system reaches a new steady state (the next message can only be triggered by an external event).

Interesting thing about modeling messages between objects: four things can happen when one object sends a message to another, and three of them are bad. First, the receiving object is not in scope and you need to change your design to ensure that it is. Second, the receiving object doesn't understand the message because it doesn't line up with one of the object's methods, so you need to change your design to fix either the sending or receiving object. Third, the pre-conditions for the receiving object are not met, so the object rejects the message and you need to change the design to make sure the pre-conditions get met. Finally, the object is there, it understands the message, and the pre-conditions are met, in which case you can assume that method's post-conditions are now true and move on to the next message. It can be kind of fun to trace messages through a design, and you get the same feeling when it works as you do when the code runs right the first time (and *that* feeling never goes away).

While this might sound like a lot of work, it is much less work than writing the code necessary to test it live, especially if you're not sure of the solution. You can change the design mathematically and take new measurements, so you can track your design goals and make valid tradeoffs as you analyze the design.

To make a long story shorter, the design worked. In fact, the whole process would run in less than an hour on a 386-based personal computer. Your smartphone has more computing power.

I reserve this technique for those design problems that are new, very difficult, or experimental. Most of the time, I go with what I've done before. You will find that 75% or more of every design is standard stuff you do on every project. If you use patterns and designs you've used before, you already know they work. Consider this another shortcut: test only those parts of a design you're not sure about, or where the risk of failure can derail the whole project.

4.15   SUMMARY

Creating a design is much like writing an outline. Creating an outline makes the writing go much more smoothly; creating a design makes the coding go much more smoothly. Despite appearances, I don't advocate doing a full-on design before you write any code. Quite the contrary, I really advocate doing just enough design, just in time, to get the code right the first time. I know it's counterintuitive, but most people seem to operate on the principle that there's never time to do it right, but always time to do it over. I'm telling you that's bunk.

If you've implemented some design pattern before and you're fairly certain it fits your problem, or some part of your current problem, by all means use the pattern. Reusing patterns saves time on the design, testing, and troubleshooting, and is one of the reasons that an experienced developer can consistently code circles around less experienced folks.

Finally, this doesn't take as long as you might think. Sure, designing something like the labor optimizer described in the last section will take time, but problems like that, along with those described in this chapter, are certified Hard Problems that demand extra attention and will kill you if they don't get it. For the most part, it takes but a few minutes to sketch out some options for a design problem and run some calculations as to which option is the best for your situation. The lesson is not to do all the design at once, but the do the design properly and effectively in those cases where it absolutely needs to be done.

4.16 FURTHER READING

Advancements in application security keep coming, Among them in Attribute-Based Access Control (ABAC) that uses attributes of the user, organization, and context to determine access. The following sources are the most accessible on the topic:

- Hur, J. and Noh, D. K. (2011, July). Attribute-based access control with efficient revocation in data outsourcing systems. *IEEE Transactions on Parallel and Distributed Systems*, 22(7):1214–1221.

- Hu, V. C., Kuhn, D. R., Ferraiolo, D. F., and Voas, J. (2015, February). Attribute-based access control. *Computer*, 48(2):85–88.

While I am not a fan of event-driven architecture, there is some good reading on the topic:

- Michelson, B. M. (2006). Event-driven architecture overview. 2(12):10–1571.

In addition to the method defined in Chapter 6 of [5], one of the leading sources on analysis and testing of software designs and architectures is:

- Clements, P., Kazman, R., and Klein, M. (2002). *Evaluating Software Architectures: Methods and Case Studies.* New York, Addison-Wesley.

CHAPTER 5

Make Your Life Easier Down the Road

Don't avoid one-time development expenses at the cost of recurring operational expenses.

–Michael T. Nygard

Modules are the units of release, and the units of reuse, which Martin calls the Reuse/Release Equivalence principle [1]. The components in a module must be released together, and because they are in the same module, they will be reused together. This leads to two additional principles of Martin's that we'll get to in the next section.

The point of this chapter can be summed up in a single quote from Nygard [6, p. 3]: "Release 1.0 is the beginning of your software's life, not the end of your project. Your quality of life after Release 1.0 depends on choices you make long before that vital milestone."

Grady Booch [10] first noted that choosing the right abstractions then organizing them into modules are two independent design decisions. Logical design is like a circuit schematic in electrical engineering in that it lays out various components of the solution and how they work together to get the job done.

Electrical engineering has a specialty known as package engineering. Packaging involves laying out the circuit on a physical circuit board (or boards) so that it fits within the confines of the space allotted to it, appropriately deals with the heat generated by the circuit, doesn't generate enough radio frequency (RF) interference to disrupt other parts of the circuit or other nearby devices, and can be maintained as necessary.

The package engineering equivalent for software it the allocation of functionality to modules. Modules may be single source files or groups of source files that are always deployed and modified together. Careful attention to the packaging of your application's functionality into modules can make your life after Release 1.0 much easier.

If you use a compiled language, such as C, C++, or Java, the perils of bad packaging become apparent every time you compile. If you've done your packaging well, most changes will affect only one or two modules, and builds will go much faster. Get this wrong, and every build will be a painful reminder.

If you use an interpreted language, such as Python or Ruby, the effects of bad packaging will be subtle bugs that show up once in a while, but not always, and are usually the result of two modules interfering with each other. Finding these can be a nightmare, fixing them can be worse.

Given the tradeoffs you have to navigate, module design can be the most difficult part of the whole project. Further, since you will likely get the allocation wrong the first try or two, it will likely change as you gain experience with where changes occur and how components get used by yourself and others. As difficult as it is, though, there are only two steps:

1. allocate design components to modules and

2. look for dependency cycles

Physically, a module can be a source file (or a header/source file pair in C or C++), a library module (dynamic link library, gem file in Ruby, a package in Python, or a .jar or .war file in Java), or a collection of files that always move together (such as you might see in a GitHub repository or Docker container on DockerHub). The format you choose depends on how you plan to distribute it. I create Python modules currently, and package them as source files that I share among several applications. Were I to contribute that code to the Open Source community, I might package it differently.

The best way to understand module dependency is to watch what your environment manager tool does the next time you install or update a new module. My current coding environment uses Anaconda Python, and the package manager, Conda, lists out the packages upon which the package I am installing depends, including specific versions. When you install or update, it will tell you which other packages will be updated, installed, or even downgraded (it does happen). Conflicting package dependency requirements create all sorts of problems that are completely unrelated to design and construction of software. Lest you roll your eyes and badmouth Python, Ruby and its gems are just as bad, and Java and Maven are worse. Good tools exist in all of these environments to make this task much less prone to error and confusion.

5.1   PICKING TEAMS

The goals of module design are to limit the scope of potential changes (isolate volatility) and maximize the ease with which the module can be released (deployed) and used apart from the rest of the application. You can accomplish both of these by maximizing module cohesion. Martin describes three forms of module cohesion that tend to fight one another, each based on one of his module design principles [1, pp. 146–147]. This tension is the main reason that Booch separates component design from module design. In this section, we start with structural dependencies between design components, ignoring for the moment message connections, to create our initial allocation to modules. Then, we apply two of Martin's three principles to adjust our allocation to optimize ease of deployment and reusability down the road.

You might be thinking that your main design components are good for only your current application. My experience says otherwise. Once you go to the trouble to capture the structure

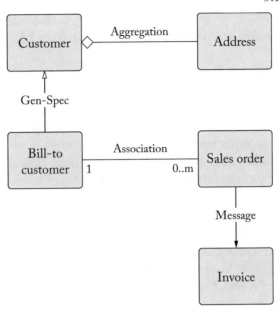

Figure 5.1: The four main types of relationships.

and behavior of an abstraction in the problem domain, you will want to reuse that work over and over. Others will, too.

5.1.1   DO I NEED TO SEE YOU?

Simply put, any connection between design components, whether structural or communication, creates a dependency that flows one way or the other. Depending on the type of connection, the dependency may cause changes in one of the abstractions to require changes in the other. Most often, these *dependent changes* follow lines of structural connection. For communication connections, a dependent change is required only when the receiving method interface or signature changes. Figure 5.1 shows examples of the three main types of structural connections and a communications connection.

Dependencies also create visibility requirements. A design component needs to be able to find a component on which it depends. Visibility has meaning both in the design and at runtime. In the design, visibility of a component is required to build components that depend on it. Visibility failures result in failed compiles in C, or immediate errors in Python as the interpreter gathers all the necessary resources. At runtime, a visibility failure means a component might not be in scope to handle its share of the processing or to receive a message which results in a defect that can be hard to find.

This step is mainly about understanding the nature and implications of each connection, since they are already (or should be) identified. You want to know, for example, whether an internal change on one end requires a dependent change on the other, or whether it might require such a change. For now, just note those connections that might create a dependent change scenario.

The lines of visibility provide a good starting point for allocating your design components to modules. Not all of the lines are created equal; two components connected by only a message link can safely reside in different modules, especially if the message interface is designed to isolate message senders from changes inside the receiver. We are mostly concerned about structural dependencies, especially gen-spec relationships and aggregations.

Associations can reside safely in separate modules depending on how you chose to implement the relationship. If you use foreign keys or other forms of reference (such as pointers in C or C++), your associated components are less tightly coupled, and less dependent, than if you chose to use internal containers. The latter case should be treated more like aggregations.

Aggregations and gen-spec participants belong in the same module, for the most part. It is reasonable to use a gen-spec relationship to inherit functionality from a library component, but only if that component is designed that way. Look out for side effects of inheriting code you didn't write. It's a useful tool, but can be very dangerous. That said, you often don't have a choice. Much of the functionality in Django, for example, is provided to your components by inheriting from Django components. Fortunately, Django components are specifically designed to be used via inheritance and side effects are minimized.

If you designed one or more walled compounds (see Section 4.5) into your set of components, each compound should go into its own module. The internals of the compound can be spread over multiple modules, but the top module that presents the compound's interface or façade to the rest of the application should fully encompass all modules below it; lower-level modules should not contain any components that are not part of the compound.

You need to watch the size of each module. Modules that are too "fat," that is, contain too many design components, will be volatile just because of their size. You can break them up into smaller modules, remembering the rule for walled compounds. When you break a module into smaller modules, they should not contain any components that were not already in the larger module. Put another way, don't go mixing your compounds; like chemistry, this can lead to unintended explosions.

5.1.2 COMPONENTS THAT CHANGE TOGETHER GO TOGETHER

Now that you have a first cut at component allocation, let's mess it up by looking at it a couple of different ways. We'll make use of two of Martin's three design principles for module cohesion: the Common Closure Principle [1, pp. 142–143] in this subsection and the Common Reuse Principle [1, pp. 144–145] in the next.

The gist of the Common Closure Principle is that all of the components in a module should be subject to the same forces for change. You change one, odds are you'll have to change another. At the very least, all of the components in the module will have to be tested together. At this point, it doesn't matter if your code is compiled or not.

As you look at your modules, you will notice groups of components that will likely, as in "might," change together. If these components span modules because of structural requirements, leave them in separate modules. If you find that a module contains components in more than one group, however, split the groups into separate modules.

Martin suggests using the "reason for change" as a criteria for these decisions, but that can lead to modules that reflect the structure of the using organization, something you really want to avoid. It's sufficient to know that a change to one component in a module will lead to changes in other components in that module; the reasons are mostly irrelevant.

5.1.3   COMPONENTS THAT ARE USED TOGETHER GO TOGETHER

Next, look at how components are used. The Common Reuse Principle says that components that are used together belong in the same module. As stated, this sets up the tug-of-war that Martin discusses in [1]. Used as a way to refine the allocation you have after the first two passes, it doesn't create the same conflicts.

The idea behind this principle is that you want to be able to reuse a design component without it dragging along a lot of other components you don't need. However, if the component you want to reuse depends on these other components, they're coming along whether you want them or not.

Again, as you examine your modules, don't recombine any modules you have. Look at each module and think about the components that will get reused together. You will likely find that some modules will contain multiple groups of components, each group reused in different contexts. These modules can be safely split.

More modules don't hurt; you can still release and reuse groups of modules. What you don't want is a module that spans more than one release or reuse group because you create a dependency between the groups in the module, effectively tying them together.

5.2   CAN I GET HERE FROM HERE?

Once you've allocated your design components to modules, note the lines of visibility between modules. These are the results of a component in one module needing visibility of a component in another module. These lines have a direction; they flow only one way. Collectively, your modules and these lines of visibility form a directed graph, a mathematical structure with some interesting properties and useful tools for analysis.

Primarily, your graph should have no cycles. A cycle is formed in a directed graph when you can start at a module and follow the directed lines linking them and eventually end up back where you started (start at Node 8 in Figure 5.2 and follow the arrows). If you've ever encountered a cycle in a large make file, you know what I'm talking about.

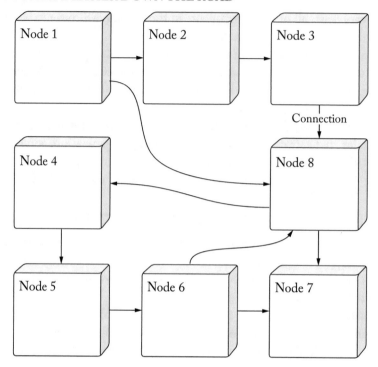

Figure 5.2: A dependency graph with a cycle.

Cyclic dependencies can kill your software. They must be avoided at all costs. If you find a cycle, you need to rework the allocation of components to modules to eliminate it. Fortunately, the links between modules are structural links between components, so removing cycles isn't as hard as it sounds. This effort is worth every minute.

Mutual importing of modules (#include in C or C++, or import in Python) doesn't necessarily create cycles. Such import statements only make the internals of one module visible to the internals of another. If a component of one module sends a message to a component in an imported module, you don't have a cycle (cyclic message connections are rather common and mostly harmless). This means that simply looking at imports is insufficient for finding cycles.

5.3   DON'T FORGET TO LOOK OUTSIDE YOUR CODE

Your software, no matter how good or comprehensive, cannot survive on its own. Even a Django or Rails application requires the support of a large number of components that aren't part of the application itself. Web servers, content servers, load balancers, database management systems, operating systems, and messaging systems are all examples of things upon which you depend but do not control. These components, as well as others in the external environment, can trip up your application when you least expect it. These components form what can be collectively referred to as "the production environment," a place to be feared, by you and your application.

5.3.1   YOUR APPLICATION IS NOT ALONE

Application software, especially when it spans a network, needs the support of a number of different kinds of servers. On the front end, load balancers help direct incoming traffic to nodes that are less stressed than others, and can even start and stop additional nodes as required by the load. Web servers, such as Apache or Nginx, render the HTML that becomes your web pages and direct incoming requests to the correct receiver in your application. Application servers, such as JBoss, Tomcat, Mongrel (for Rails) or Gunicorn (for Django), work with your application to handle traffic to and from and within your application.

In the middle, the operating systems for the servers you use bring with them support for creating and executing programs. If you use a compiled language, the compiler and run-time environment for your code can change with a simple operating system upgrade. If you use an interpreted language, your runtime environment is even more fragile, as an upgrade to any library can immediately break your code. To combat this problem, I have taken to creating Docker containers to host my applications. I create a Docker image that contains the operating environment that I want, including any front end servers, and then deploy those containers. Deploying my applications to a new server is a matter of copying the container image to the server, creating the container, and then starting it up. Aside from the time to copy the image, the process takes about five minutes. For me, a typical deployment consists of a single container for each application plus one additional container for the database, which may or may not be on a different server. More sophisticated operations use tools like Kubernetes to automatically deploy and manage collections of containers.

On the back end, you need to store your data somewhere, and most applications use some combination of a database management system and a file structure. Most database management systems are based on the relational data model, but other models are out there, including network and semantic. Some applications, particularly those with large transaction loads and mixed forms of data, use some form of NoSQL data store, such as Hadoop.

Before you can test your application, you have to figure out which components you need and which specific products you plan to use. Then, you have to learn how to install, configure, and operate them. No longer can you be "just" a programmer, you have to be part sysadmin to get anything done. As an aside, those of us who tend to work on our own have *always* followed the DevOps model, and frankly, any application fielded today requires some flavor of DevOps.

5.3.2   REDUNDANT GEOGRAPHY

Stuff happens. Even the best-designed systems go down. Netflix has suffered several outages, but the two big ones happened because of a problem in the Amazon AWS data center in which they were running. In Chapter 2, we discussed the ways to achieve high availability, expressed as percent uptime and measured in terms of the number of nines in the value. To go from 4 nines (99.99% uptime) to 5 nines or higher requires redundant, well, everything. The pattern for achieving five nines, which Michael Nygard calls Bulkheads [6, pp. 108–112], includes:

- multiple active databases (active-active),

- multiple load balancers,

- multiple server clusters,

- multiple instance of the application,

- multiple data centers, and

- multiple climates.

In short, you want to find and remove every single point of failure so that when stuff happens, only the portion of the system directly affected goes down while the rest of the system keeps running. Points of failure can be sneaky and hard to find. It can take a fair bit of creativity to recognize one. Sometimes, when you find and fix one, you find you have just moved it or created a new one. After every change to the design, you have to analyze the entire design again, including all of the external environments.

Infrastructure folks have lots of ways to implement redundancy, and if you find that is beyond your skill set, as it is for most of us, get with a skilled infrastructure architect. My first project as an enterprise architect at T-Mobile was to design the infrastructure for the new data center. I knew enough about data center design and operations to be dangerous. Fortunately, a colleague and I were able to work with a couple of top-notch data center consultants. The results of that project was a well-documented design Operations could use, and two much smarter enterprise architects. Most of the techniques we learned are still in use. The most amazing thing I learned is that *everything* can be redundant.

Implementing redundancy requires a lot of work and generally costs a lot. You have to weigh that against what it costs to be down. When I worked at Nordstrom in the mid-2010s, we calculated that every hour of downtime resulted in $150,000 in lost sales for the web store (it's a lot more these days). It doesn't take a lot of downtime to pay for a whole lot of redundancy.

5.3.3 POOL CONNECTIONS

Opening a database connection is nearly as costly as creating a thread or process. Database connections, threads, and processes are resources that can be used repeatedly by an application. Creating a pool of such resources of the appropriate size that are left open or allocated, and ensuring that the pool size can be changed, can dramatically speed up the perceived performance of an application.

The first version of the case study application was developed using a code generator tool built for the IBM AS400 environment, later adapted to an SQL environment. We didn't discover what that meant until we'd been in production for a while, when it was too late to fix. Every access to the database created and opened a connection, which includes opening a socket to the database server, authenticating the user, and setting up the database session. After the data was retrieved or updated, using dynamically generated SQL, the connection was torn down. An experienced database administrator or programmer should recognize the many red flags in the preceding description.

Judicious use and management of a pool of connections [6, pp. 197–198] that are initialized at startup can solve most of these problems (not much you can do about a tool that insists on doing everything with dynamically generated SQL). When a connection goes bad, as they sometimes do, you can close it and open a new one. When load increases, new connections can be opened and added to the pool, and connections can be closed as load decreases. Connection pool management is supported by most languages and every database management system, so there is really no excuse to not take the extra effort that will repeatedly save you, and your users, time.

5.3.4   CACHE CONTENT

Caching [6, pp. 199–200] is a means of storing retrieved data in memory so it can be used repeatedly without spending the resources to retrieve it every time. Caching can be very simple, loading an internal variable with a value retrieved from the database, to very complex, using a combination of memory and disk to store the cache.

Many languages provide support for caching, usually through 3rd party libraries. In my Python applications, I use the memcache package to cache whole web pages, which has the added benefit of being very easy to use with Django. The library provides the ability to set the maximum size of the cache as well as a time limit, forcing a new retrieval after it expires. The main page for one of my applications contains data from nearly 50 separate SQL statements. Executing these takes 5–10 seconds, which isn't bad, until you suffer through that every time you hit the main page. The cache in that case is set to expire every 48 hours.

Stale data is the bane of caches. Data becomes stale when the value behind the cache changes but the cache is not updated. If you decided to use a cache, make sure to include a mechanism to refresh it after a period of time or when the data changes. To be honest, I don't find the need to use caching very much, but applications with high user loads and a high proportion of fairly static content make caching a requirement.

5.3.5   PUSH CONTENT TO THE EDGE

A *content delivery network* (CDN) is a set of caching servers that are positioned on the edges of a network, distributed geographically to be close to users. A CDN is a sophisticated form of caching, with all of the benefits and risks. Amazon uses multiple CDNs for catalog data and prices. In many cases, they use a CDN to store product quantities, but discovered that works only when you have lots of an item on hand. Significant effort happens behind the scenes to monitor each CDN and push out updated data when necessary.

Amazon is an extreme case calling for extreme solutions. Most application don't have the need for that level of sophistication, and opt to merely store static HTML and stylesheets on a CDN, if a CDN is used at all.

In a slightly less elegant example of a CDN, one of my applications maintains a file structure that hold about 4 TB (yes, that's terabytes with a "T") of images. There is currently a copy of that structure in two locations. We maintain two copies because of the need for users of the application to access the data from more than one location. Most of our users access the application via the cloud which uses a copy maintained in the cloud, while those of us in the lab access the local copy. If you've ever tried pulling a 5 GB image across the Internet, you know why we added the local copy of the stash. We synchronize the two copies using background jobs that run nightly.

5.3.6   PRECOMPUTE VALUES

Nygard [6, pp. 201–204] discusses this pattern as a way to pre-render HTML to save time. What he describes is really a form of caching. Precomputing values and caching the result often go together, so don't get into any arguments about whether you are caching or precomputing; odds are you're doing both.

You can precompute values by doing calculations and storing the results in a database even if the data for the calculation is readily available. In the case study application, we computed the line extensions by multiplying the quantity by the price and storing the result on the line. If a currency conversion was required, we did those at fixed points in the state of a transaction (when an outbound order was shipped, or when we received notice of an inbound shipment being on the way) to line up with the business event.

One of my applications involves using information from multiple images to compute parameters for a tumor growth model. Until recently, that process was expensive and time consuming, so we computed those values as soon as the data was available and stored it in a table in the database. We don't do that anymore for a number of reasons, but it remains a good example of precomputing values.

5.3.7   CREATE STEADY STATES

Nygard's description of the Steady State pattern [6, pp. 113–119] is really about automatic maintenance of mechanisms that consume resources, such as log files, database rows, or data caches. We talk about deleting database entries and data caches elsewhere, so we'll focus on log files here.

The point of the pattern is that log files and other resource consumers needs to be maintained. You can log onto the server periodically and delete files, or you can create a mechanism that will do that for you. I use cron to schedule shell scripts that maintain these log files. They run monthly which is often enough for my purposes. Nygard says they should run more often, and that data that is too old (more than a week) is no longer needed and should be discarded. Setting up cron jobs is simple, and is one of the reasons I deploy my applications on Docker containers because they can all run on Linux installations, even on a Windows machine.

The principle here is that you want to minimize manual intervention in a running application as much as you can. Think "fire and forget." You fire the application up, then forget about it until it tells you it needs attention.

5.3.8   DATABASE REPLICATION

If your application's user base is large or distributed enough, you will end up with multiple databases that have to be kept in synch with each other. Even if all you have is a backup database for redundancy, you need a way to keep it up to date. Nearly all database management systems provide several techniques that work behind the scenes to make this happen. You can also build it into your application. The route you take depends on a lot of things.

One of the quirkier features of every relational database system I've worked with, which is very nearly all of them, is that the normal read process starts by locking the table against changes. You can either read uncommitted data (a "dirty" read), or you lock the table. With a single database, under normal load conditions, this isn't usually a problem.

On the project that inspired the case study, our plan was to create 40 distributed databases shared by 70 distribution centers. This pattern matched the data generation and use pattern discussed in Section 2.3.4, as the vast majority of data usage was confined to the local distributed copy. (It had also been over 10 years since that discussion about minicomputers vs. mainframes, and we didn't use either one very often anymore). A distribution center could sell out of the inventory of another if the costs to ship the goods didn't exceed the value of the order. This action was also most often confined to the local distributed database, but not always.

In addition, all transactions were to be replicated to a central database that was used for reporting and back-office functions, such as updating the general ledger, product definition, managing customer credit, and pricing.

Based on the advice of our database vendor (it doesn't matter who, they're all the same), we implemented transactional replication in which every transaction was transmitted to the central database. All transactions made directly in the central database were pushed out to the distributed copies for use locally. After the second distributed location was brought online, we found that the central database spent over half of the time with tables locked for replication. We discovered this when processing came to a screeching halt and we saw the page locks on indexes in the logs; we immediately understood why database administrators typically drop all indexes before a big data load and then rebuild them afterward (this wasn't an option for us; we looked at it carefully).

We were forced to change replication schemes between rolling out the second remote site and the third, all while mainlining the production capability for the first two locations. It was a lot like changing the engine on a plane while in flight.

In the end, we adopted a log-based replication scheme that transmitted log entries as they were written to be applied by the receiving databases in the order received, even if they got backed up, sometimes by as much as five minutes.

The point of this story is when you plan to use multiple databases, you need to be deliberate about *how* those databases synch themselves, even if the database software does it for you. In that particular application, the key measure was the number of database hits per transaction, and we had to work hard to minimize them. At the time that system was built, there were no easy ways to cache content in a network, and we built everything from scratch.

5.3.9   AUTO-GENERATED DATABASE KEYS

Another side-effect of using multiple active databases is that each instance is generating database keys for new rows. You will inevitably have to deal with duplicate database keys if you try to combine this data as we did in the case study. You can generate your own keys, but this is almost never worth the extra work required.

Typically, a database runs an internal counter for the next key for every table that has an auto-generated key. All new databases will start with 1 for each table, which means that by the time you have 40 databases installed and generating keys, you will have 40 copies of the same key value pointing to 40 different rows, for the first several thousand rows in each database. Think of multiple thousands of duplicate database keys trying to mingle in the central database, which generates its own keys.

Our chosen solution was unique, in that I don't know of anyone else who has used it. We assigned each database a unique identifier when we installed it. This database ID was then combined with the row ID generated by the database instance to create the full key for the row. This worked in our case because we controlled the creation of database instances and their identifiers.

Amazon Web Services (AWS) includes an auto-scaling feature which allows additional instances of computing resources, including databases, to be created on the fly as load demands. This means that you don't control the creation of database instances and our solution would no longer work. AWS may have a solution for duplicate database keys, but I have yet to find it.

Like replication, this is a fundamental design issue that needs to be solved before the first instance is built. Failure to do so deliberately will result in failure of the system, but not until it goes into production. Many stories of failures of new websites going live could be attributed to this kind of problem. People know how to build the front end for high volume, but few even think about the back end.

5.3.10   CONSIDER MACHINE LEARNING

In 1959, A. L. Samuel [33] wrote that machine learning is a "Field of study that gives computers the ability to learn without being explicitly programmed." The idea is that machines learn to do a task on their own, rather than being programmed to do it. "Programming" a computer using machine learning techniques means training a data structure and program, such as an artificial neural network, to perform a task, such as sorting members of a data set into a set of classifications. These classes can be predetermined, or you can let the training process determine them. The classes can then be used to make decisions about how to perform a task, such as whether to approve a loan application. The basis for this learning is by example and experience, or trial and error [34]. Once trained, the resulting program is typically called a "model."

The implication is that any problem that can be solved by following examples or trial and error might lend itself to a an application of machine learning. In my research lab at the Mayo Clinic, classifying magnetic resonance images into their sequence types was a complex, difficult task that took many hours to master. We trained an artificial neural network to classify images and found that the machine model was much faster and more accurate than an experienced human [35].

There are three basic methods for training computers [36]: supervised, unsupervised, and semi-supervised. We'll cover each of these methods briefly in the following subsections.

Supervised Learning

Supervised learning starts by exposing your model to examples known to be correct. These examples are said to be "labeled" with the correct classification, which is called the "ground truth." In our research lab application, we had access to thousands of images that had been correctly classified which we used to train our network. In the process, the training program makes predictions using the input data, compares those predictions to the ground truth, refines its internal parameters and tries again, repeating until the desired level of accuracy is achieved. The "distance" between the prediction and ground truth is measured using a cost function that is minimized over the course of the training.

Supervised learning is used in many fields including biology and medicine. The ability to use supervised learning depends on the availability of a large pool of diverse data, including some counterexamples (examples incorrectly labeled, but known to be incorrectly labeled). The larger and more diverse the pool of training data, the better. Lack of diversity, or biases embedded in the data, leads to biases in the resulting model. The biases can range from hidden to annoying to downright harmful.

Shanthamallu et al. [34] provide many references for the details about specific supervised learning techniques.

Unsupervised Learning

In unsupervised learning, the model is given a set of data and asked to figure out the classifications and patterns on its own [37]. There are no labels of ground truth, nor are there labels of what is known to be not true.

The most common application of unsupervised learning is to classify a data set into some number of classes or clusters, so that the data elements within a cluster are similar, but are dissimilar to data elements in other clusters. The training algorithm determines the clusters and the criteria for judging similarity.

Unsupervised learning has applications in many disciplines, including medicine, business decisions, ecommerce, and fault detection in manufacturing [34].

Like supervised learning, the results obtained from unsupervised learning is dependent on the size and diversity of the training data, only more so. It can be the case that a model trained on one set of data completely fails when applied to another set. This should not be a surprise to those of us who have had the same experience ("I'm not sure what to do here, I've not seen this case before.").

Shanthamallu et al. [34] provide many references for various unsupervised learning techniques.

Semi-Supervised Learning

In semi-supervised learning, a common condition is that some of the positive results in the data set are labeled, that is, you know some elements to be labeled correctly, but there are no counterexamples (no negative or incorrect labels), and the vast majority of the data is unlabeled. This is known as a Positive and Unlabeled (PU) problem [38]. Jaskie and Spanias [38] describe several applications of the PU learning problem, including recognition of objects in satellite images, identifying tumors on medical images, recommendations for business ecommerce applications, and some security and signal processing applications.

Jaskie and Spanias describe four main approaches to PU learning, including an emergent method called generative adversarial networks (GANs) [39]. GANs are interesting, and troubling, enough, that we need to discuss them briefly.

In a GAN, two networks are trained: one to generate the predictions, and the other to determine whether the predictions are incorrect. The objective is for the first network to become good enough to fool the second some predetermined percentage of the time.

GANs are a powerful tool, and have recently been used to produce faked images of people that are so good, it is difficult for a human to tell the difference. Applied one frame at a time, it is possible to produce a video of anyone saying or doing anything, and make it nearly impossible to tell the video is fake.

5.3.11 MAKE ROOM FOR MATLAB®

MATLAB is basically an integrated development environment (IDE) with a proprietary language designed to make matrix and numeric analysis much easier than using other languages. MATLAB is short for MATrix LABoratory [40], and provides a lot of support for matrix programming. MATLAB is popular among scientists and engineers because of this. The MATLAB language closely resembles Python, to the point that porting MATLAB code to Python is fairly easy to do, and is becoming more common in both university and commercial settings. Many of the features that are part of the application at my research lab were originally written in MATLAB.

Stephen J. Chapman [40] lists several advantages that MATLAB has over general-purpose languages, including Python. Among them are the richness of the built-in functions, including single functions for solving many complex mathematical tasks, and the fact that it runs on nearly any computing platform. MATLAB also provides device-independent plotting, allowing for rather sophisticated data visualization. These advantages are real, especially when you need to do a quick analysis of a structure or manipulate some matrices. The main disadvantage is the cost, at many thousands of dollars per copy. Keeping a local copy of MATLAB up to date is no picnic, either.

MATLAB has been used in many applications, including machine learning [41, 42], image processing [43, 44], Kalman filter estimation [45], the Internet of Things [46, 47], and speech and audio compression [48–50]. The sources listed provide details into the various implementations to more easily make use of MATLAB to solve specific problems.

The advent of Open Source libraries such as SciPy [51], NumPy [52], Matplotlib [53], and others, including at least one that allows Python to open, navigate, and read MATLAB files, narrows the advantages that MATLAB holds over other languages and are the main driver behind the trend to port MATLAB code to Python.

5.4 SUMMARY

You application lives in a hostile environment. Your application will change over time, and the environment will change under it. Paying attention to how you package your application for deployment can make your life easier down the road.

To be useful, every application needs the support of any number of additional tools and services. Selecting, installing, and managing these services is part of engineering your application, so approach it with the same care and deliberation you use for the design itself.

In addition, there are things you can do with your application to help it perform better once it is in the wild. Several patterns were described that can be used together or independently to improve the perceived performance of your application; with performance, perception is everything.

Finally, consider alternatives to writing your own code, such as machine learning or using MATLAB if the tools happen to fit your situation.

5.5   FURTHER READING

My original source for large-scale packaging design was this paper from John Lakos, which was later turned into parts of several books, one of which is also provided:

- Lakos, J. S. (1992). Desinging-in quality for large C++ projects. *Proc. of the Pacific Northwest Software Quality Conference*. pp. 275–284, Portland, OR, PNWSQC.

- Lakos, J. S. (1996). *Large-Scale C++ Software Design*. Englewood Cliffs, NJ, Addison-Wesley.

Here are a couple of sources on content delivery networks (CDN) for pushing content to the edges:

- Saroiu, S., Gummadi, K. P., Dunn, R. J., Gribble, S. D., and Levy, H. M. (2002). An analysis of internet content delivery systems. *ACM SIGOPS Operating Systems Review*, 36(SI):315–327.

- Mulerikkal, J. P. and Khalil, I. (2007). An architecture for distributed content delivery network. *15th IEEE International Conference on Networks*, pp. 359–364.

The following sources provide more detailed information on techniques for training machine learning models:

- Shanthamallu, U. S., Spanias, A., Tepedelenlioglu, C., and Stanley, M. (2017). A brief survey of machine learning methods and their sensor and IoT applications. *8th International Conference on Information, Intelligence, Systems and Applications (IISA)*, pp. 1–8, IEEE.

- Jaskie, K. and Spanias, A. (2019). Positive and unlabeled learning algorithms and applications: A survey. *10th International Conference on Information, Intelligence, Systems and Applications (IISA)*, pp. 1–8, IEEE.

The following sources discuss various uses for machine learning algorithms, but not necessarily training them:

- Samuel, A. L. (1959, July). Some studies in machine learning using the game of checkers. *IBM Journal of R&D*, 3(3):210–229.

- Libbrecht, M. W. and Noble, W. S. (2015). Machine learning applications in genetics and genomics. *Nature Reviews Genetics*, 16(6):321–332.

- Kourou, K., Exarchos, T. P., Exarchos, K. P., Karamouzis, M. V., and Fotiadis, D. I. (2015). Machine learning applications in cancer prognosis and prediction. *Computational and Structural Biotechnology Journal*, 13:8–17.

- Gao, J. (2014). *Machine Learning Applications for Data Center Optimization*. Google Research.

CHAPTER 6

Conclusion

Creating high quality software that solves a business problem isn't easy, but it isn't as difficult as it seems. By adding some engineering discipline to your design efforts, you can be reasonably sure that the solution you produce is the optimum solution for your situation.

Agile purists continue to argue that design is not necessary, and go so far as to eliminate it from the schedule. Despite their best effort, design happens, and when it does, engineering should happen too. Design doesn't have to happen all at once, and not all components need to be engineered.

Applying the techniques in this book will lead to better software, as they have for me over the last 25 years (it took the first 15 years to make enough mistakes to learn them). My aim for sharing these techniques is to encourage their use by folks who need to write software as part of their job, but who are not professional developers.

As more part-time developers apply these techniques, the overall quality of software in various fields will improve. Remember, no software is one-time only; it will always last longer than you expect, and if you don't make the effort when you first create it, you will regret it every time you use it after that, and those who follow you will curse you under their breaths, or maybe out loud. You just have to trust me on that.

Happy Engineering!

Bibliography

[1] R. C. Martin, *Clean Architecture: A Craftsman's Guide to Software Structure and Design*, p. 58, Prentice Hall, New York, 2018. 1, 5, 7, 8, 37, 39, 61, 95, 96, 98, 99

[2] J. Mostow, Towards better models of the design process, *AI Magazine*, 6(1), 1985. 1

[3] H. Petroski, *To Engineer is Human*, St. Martin's Press, New York, 1985. 2, 4, 20

[4] S. A. Whitmire, Tools may come, and tools may go...but some things never change, *Pacific Northwest Software Quality Conference Proceedings*, Portland, OR, 2004. 2

[5] S. A. Whitmire, *Object Oriented Design Measurement*, John Wiley & Sons, Inc., New York, 1997. 2, 10, 11, 12, 15, 16, 18, 21, 39, 49, 55, 59, 60, 65, 66, 68, 92, 94

[6] M. T. Nygard, *Release It! Design and Deploy Production-Ready Software*, The Pragmatic Programmers, Raleigh, NC, 2007. 4, 30, 33, 34, 66, 95, 102, 103, 104

[7] B. Meyer, *Object-Oriented Software Construction*, Prentice Hall, Englewood Cliffs, NJ, 1988. 5

[8] B. Liskov, Data abstraction and hierarchy, *Addendum to Proceedings of OOPSLA*, Orlando, FL, 1987. DOI: 10.1145/62139.62141. 6

[9] E. Gamma, R. Helm, R. Johnson, and J. Vlissides, *Design Patterns: Elements of Reusuable Object-Oriented Software*, Addison-Wesley, Reading, MA, 1995. 7, 13, 17, 36, 73, 74, 78, 79, 80, 81

[10] G. Booch, *Object-Oriented Analysis and Design with Applications*, Benjamin/Cummings, Redwood City, CA, 1994. 11, 15, 54, 95

[11] G. Myers, *Software Reliability: Principles and Practices*, John Wiley & Sons, New York, 1976. 14

[12] E. V. Berard, *Essays on Object-Oriented Software Engineering*, vol. 1, Prentice Hall, Englewood Cliffs, NJ, 1993. 15

[13] M. Fowler, *Patterns of Enterprise Application Architecture*, Addison-Wesley, New York, 2003. 19, 20, 31, 83

[14] F. Buschmann, R. Meunier, H. Rohnert, P. Sommerlad, and M. Stal, *Pattern-Oriented Software Architecture: A System of Patterns*, John Wiley & Sons, Inc., New York, 1996. 19, 28, 29, 31, 36

[15] J. Woodcock and M. Loomes, *Software Engineering Mathematics*, Addison-Wesley, Reading, MA, 1988.

[16] S. Stepney, R. Barden, and D. Cooper, *Object Orientation in Z*, Springer-Verlag, London, 1992. DOI: 10.1007/978-1-4471-3552-4.

[17] C. Alexander, *A Timeless Way of Building*, Oxford University Press, New York, 1979. 25

[18] T. DeMarco, *Controlling Software Projects*, Yourdon Press, Englewood Cliffs, NJ, 1982. 26

[19] S. A. Whitmire, 3D function points: Scientific and real-time extensions to function points, *Proc. of the Pacific Northwest Software Quality Conference*, Portland, OR, 1992. 27

[20] S. A. Whitmire, Thoughts on Solution Architecture, 2014. https://scottwhitmire@wordpress.com. [Accessed December 2020]. 27

[21] D. Ferraiolo and R. Kuhn, Role-based access controls, *Proc. of the 15th National Computer Security Conference*, Baltimore, MD, 1992. 37, 67

[22] M. Shaw, Comparing architectural design styles, *IEEE Software*, 12(6), November 1995. DOI: 10.1109/52.469758. 39

[23] R. C. Sharble and S. S. Cohen, The object-oriented brewery, *ACM SIGSOFT Software Engineering Notes*, 18(2):60–73, April 1993. DOI: 10.1145/159420.155839. 39

[24] M. Bunge, *Treatise on Basic Philosophy I: The Furniture of the World*, Dordrecht, Riedel, 1977. DOI: 10.1007/978-94-010-9924-0. 45

[25] G. Yuan, A depth-first process model for object-oriented development with improved OOA/OOD notations, *Report on Object-Oriented Analysis and Design*, 2(1):23–37, May/June 1995. 46, 54

[26] P. Coad and E. Yourdon, *Object-Oriented Analysis*, Yourdon Press/Prentice Hall, Englewood Cliffs, NJ, 1991. 54

[27] J. Rumbaugh, M. Blaha, W. Premerlani, F. Eddy, and W. Lorensen, *Object-Oriented Modeling and Design*, Prentice Hall, Englewood Cliffs, NJ, 1991. 54

[28] I. Jacobson, M. Christerson, P. Jonsson, and G. Övergaard, *Object-Oriented Software Engineering: A Use Case Driven Approach*, Addison-Wesley, Reading, MA, 1992. 55

[29] S. M. Smith, KERI_WP_2.x.web.pdf, July 21, 2020. https://github.com/SmithSamuelM/Papers/blob/master/whitepapers/KERI_WP_2.x.web.pdf 71

[30] W. W. Stead, W. E. Hammond, and M. J. Straube, A chartless record—Is it adequate?, *Journal of Medical Systems*, 7:103–109, April 1983. https://doi.org/10.1007/BF00995117 DOI: 10.1007/bf00995117. 75

[31] P. Coad, D. North, and M. Mayfield, *Object Models: Strategies, Patterns, and Applications*, Yourdon Press, Upper Saddle River, NJ, 1997. 86

[32] R. Wirfs-Brock, B. Wilkerson, and L. Wiener, *Designing Object-Oriented Software*, Prentice Hall, Englewood Cliffs, NJ, 1990. 92

[33] A. L. Samuel, Some studies in machine learning using the game of checkers, *IBM Journal of R&D*, 3(3):210–229, July 1959. DOI: 10.1147/rd.33.0210. 107

[34] U. S. Shanthamallu, A. Spanias, C. Tepedelenlioglu, and M. Stanley, A brief survey of machine learning methods and their sensor and IoT applications, *8th International Conference on Information, Intelligence, Systems and Applications (IISA)*, 2017. DOI: 10.1109/iisa.2017.8316459. 107, 108

[35] S. Ranjbar, K. W. Singleton, P. R. Jackson, C. R. Rickertsen, S. A. Whitmire, K. R. Clark-Swanson, J. R. Mitchell, K. R. Swanson, and L. S. Hu, A deep convolutional neural network for annotation of magnetic resonance imaging sequence type, *Journal of Digital Imaging*, 33(2):439–446, 2020. DOI: 10.1007/s10278-019-00282-4. 107

[36] J. L. Berral-García, A quick view on current techniques and machine learning algorithms for big data analytics, *ICTON Trento*, pages 1–4, 2016. DOI: 10.1109/icton.2016.7550517. 107

[37] M. E. Celebi, K. Aydin (Ed.), *Unsupervised Learning Algorithms*, Springer International Publishing, Switzerland, 2016. 108

[38] K. Jaskie and A. Spanias, Positive and unlabeled learning algorithms and applications: A survey, *10th International Conference on Information, Intelligence, Systems and Applications (IISA)*, 2019. DOI: 10.1109/iisa.2019.8900698. 108

[39] M. Hou and B. Chaib-Draa, Generative adversarial positive-unlabeled learning, *IEEE IJCAI*, Stockholm, 2018. DOI: 10.24963/ijcai.2018/312. 108

[40] S. J. Chapman, *MATLAB Programming for Engineers*, Cengage Learning, Boston, MA, 2016. 109

[41] M. Paluszek and S. Thomas, *MATLAB Machine Learning*, Apress, Berkeley, CA, 2016. 109

[42] P. Kim, *Matlab Deep Learning: With Machine Learning, Neural Networks and Artificial Intelligence*, Apress, Berkeley, CA, 2017. 109

[43] C. Solomon and T. Breckon, *Fundamentals of Digital Image Processing: A Practical Approach with Examples in Matlab*, John Wiley & Sons, New York, 2011. DOI: 10.1002/9780470689776. 109

[44] T. Chaira and A. K. Ray, *Fuzzy Image Processing and Applications with MATLAB*, CRC Press, 2017. DOI: 10.1201/b15853. 109

[45] M. S. Grewal and A. P. Andrews, *Kalman Filtering: Theory and Practice with MATLAB*, John Wiley & Sons, New York, 2014. DOI: 10.1002/9781118984987. 109

[46] M. Stanley and J. Lee, Sensor analysis for the internet of things, *Synthesis Lectures on Algorithms and Software in Engineering*, 9(1):1–137, 2018. DOI: 10.2200/s00827ed1v01201802ase017. 109

[47] S. Pasha, ThingSpeak based sensing and monitoring system for IoT with MATLAB analysis, *International Journal of New Technology and Research (IJNTR)*, 2(6):19–23, 2016. 109

[48] A. Spanias, Speech coding: A tutorial review, *Proc. of the IEEE*, 82(10):1541–1582, 1994. DOI: 10.1109/5.326413. 109

[49] P. Hill, *Audio and Speech Processing with MATLAB*, CRC Press, 2018. DOI: 10.1201/9780429444067. 109

[50] J. J. Thiagarajan and A. S. Spanias, Analysis of the MPEG-1 layer III (MP3) algorithm using MATLAB, *Synthesis Lectures on Algorithms and Software in Engineering*, 3(3):1–129, 2011. DOI: 10.2200/s00382ed1v01y201110ase009. 109

[51] P. Virtanen, R. Gommers, T. E. Oliphant, M. Haberland, T. Reddy, D. Cournapeau, E. Burovski, P. Peterson, W. Weckesser, J. Bright, S. J. van der Walt, M. Brett, J. Wilson, K. J. Millman, and May, SciPy 1.0: Fundamental algorithms for scientific computing in Python, *Nature Methods*, 17:261–272, 2020. DOI: 10.1038/s41592-019-0686-2. 109

[52] K. J. Millman, S. J. van der Walt, R. Gommers, P. Virtanen, D. Cournapeau, E. Wieser, J. Taylor, S. Berg, N. J. Smith, R. Kern, M. Picus, S. Hoyer, M. H. van Kerkwijk, M. Brett, and Haldane, Array programming with NumPy, *Nature*, 585:357–362, 2020. DOI: 10.1038/s41586-020-2649-2. 109

[53] J. Hunter, Matplotlib: A 2D graphics environment, *Computing in Science and Engineering*, 9:90–95, 2007. DOI: 10.1109/mcse.2007.55. 109

[54] T. DeMarco, *Controlling Software Projects*, Yourdon Press, Englewood Cliffs, NJ, 1982.

[55] K. E. Wiegers, *Software Requirements*, Microsoft Press, Redmond, WA, 2003.

[56] D. Schmidt, M. Stal, H. Rohnert, and F. Buschmann, *Patterns-Oriented Software Architecture: Patterns for Concurrent and Networked Objects*, vol. 2, John Wiliey & Sons, New York, 2000.

[57] S. Saroiu, K. P. Gummadi, R. J. Dunn, S. D. Gribble, and H. M. Levy, An analysis of internet content delivery systems, *ACM SIGOPS Operating Systems Review*, 36(SI):315–327, 2002. DOI: 10.1145/844128.844158.

[58] N. Rosanski and E. Woods, *Software Systems Architecture: Working with Stakeholders Using Viewpoints and Perspectives*, Addison-Wesley, Upper Saddle River, NJ, 2012.

[59] R. S. Pressman, *Software Engineering: A Practitioner's Approach*, McGraw Hill Higher Education, New York, 2005.

[60] H. Petroski, *Success Through Failure: The Paradox of Design*, Princeton University Press, Princeton, NJ, 2006.

[61] H. Petroski, *Small Things Considered: Why There is No Perfect Design*, Vintage Books, New York, 2003.

[62] H. Petroski, *Invention by Design: How Engineers Get from Thought to Thing*, Harvard University Press, Cambridge, MA, 1996.

[63] H. Petroski, *Design Paradigms: Cast Histories of Error and Judgement in Engineering*, Cambridge University Press, Cambridge, MA, 1994. DOI: 10.1017/CBO9780511805073.

[64] M. Page-Jones, *The Practical Guide to Structured Systems Design*, Yourdon Press, Englewood Cliffs, NJ, 1980.

[65] J. P. Mulerikkal and I. Khalil, An architecture for distributed content delivery network, *15th IEEE International Conference on Networks*, 2007. DOI: 10.1109/icon.2007.4444113.

[66] R. E. Miller, *The Quest for Software Requirements*, Maven Mark Books, Milwaukee, WI, 2009.

[67] B. M. Michelson, Event-driven architecture overview, customers.com, 2(12):10–1571, 2006. DOI: 10.1571/bda2-2-06cc.

[68] S. M. McMenamin and J. F. Palmer, *Essential Systems Analysis*, Yourdon Press, Englewood Cliffs, NJ, 1984.

[69] M. W. Libbrecht and W. S. Noble, Machine learning applications in genetics and genomics, *Nature Reviews Genetics*, 16(6):321–332, 2015. DOI: 10.1038/nrg3920.

[70] R. J. Leach, *An Introduction to Software Engineering*, CRC Press, New York, 2000. DOI: 10.1201/9781315371665.

[71] J. S. Lakos, *Large-Scale C++ Software Design*, Addison-Wesley, Englewood Cliffs, NJ, 1996.

[72] J. S. Lakos, Desinging-in quality for large C++ projects, *Proc. of the Pacific Northwest Software Quality Conference*, Portland, OR, 1992.

[73] K. Kourou, T. P. Exarchos, K. P. Exarchos, M. V. Karamouzis, and D. I. Fotiadis, Machine learning applications in cancer prognosis and prediction, *Computational and Structural Biotechnology Journal*, 13:8–17, 2015. DOI: 10.1016/j.csbj.2014.11.005.

[74] M. Kircher and P. Jain, *Pattern-Oriented Software Architecture: Patterns for Resource Management*, vol. 3, John Wiley & Sons, Hoboken, NJ, 2004. DOI: 10.1109/wicsa.2007.32.

[75] J. Hur and D. K. Noh, Attribute-based access control with efficient revocation in data outsourcing systems, *IEEE Transactions on Parallel and Distributed Systems*, 22(7):1214–1221, July 2011. DOI: 10.1109/tpds.2010.203.

[76] W. S. Humphries, *A Discipline for Software Engineering*, Addison-Wesley, Reading, MA, 1995.

[77] D. W. Hubbard, *How to Measure Anything*, John Wiley & Sons, Hoboken, NJ, 2010.

[78] V. C. Hu, D. R. Kuhn, D. F. Ferraiolo, and J. Voas, Attribute-based access control, *Computer*, 48(2):85–88, February 2015. DOI: 10.1109/mc.2015.33.

[79] B. Henerson-Sellers, *Object-Oriented Software Metrics: Measures of Complexity*, Prentice Hall, Upper Saddle River, NJ, 1996.

[80] R. Glass, *Software Conflict*, Yourdon Press, Englewood Cliffs, NJ, 1991.

[81] D. C. Gause and G. M. Weinberg, *Exploring Requirements: Quality Before Design*, Dorset House, New York, 1989.

[82] J. Gao, *Machine Learning Applications for Data Center Optimization*, Google Research, 2014.

[83] M. Fowler, *Analysis Patterns: Reusable Object Models*, Addison-Wesley, Reading, MA, 1997.

[84] N. E. Fenton and S. L. Pfleeger, *Software Metrics: A Rigorous and Practical Approach*, International Thomson Computer Press, New York, 1997. DOI: 10.1201/b17461.

[85] D. W. Embley, B. D. Kurtz, and S. N. Woodfield, *Object-Oriented Systems Analysis: A Model-Driven Approach*, Yourdon Press, Englewood Cliffs, NJ, 1992.

[86] T. DeMarco, *Structured Analysis and Specification*, Yourdon, Inc., New York, 1978.

[87] A. M. Davis, *Software Requirements: Analysis and Specification*, Prentice Hall, Englewood Cliffs, NJ, 1990.

[88] P. Clements, R. Kazman, and M. Klein, *Evaluating Software Architectures: Methods and Case Studies*, Addison-Wesley, New York, 2002.

[89] F. Buschmann, K. Henney, and D. C. Schmidt, *Pattern-Oriented Software Architecture: On Patterns and Pattern Languages*, vol. 5, John Wiley & Sons, Hoboken, NJ, 2007.

[90] F. Buschmann, K. Henney, and D. C. Schmidt, *Pattern-Oriented Software Architecture: A Pattern Language for Distributed Computing*, vol. 4, John Wiley & Sons, Hoboken, NJ, 2007.

[91] F. Brooks, *The Mythical Man-Month*, Addison-Wesley, Reading, MA, 1975.

[92] B. W. Boehm, *Software Engineering Economics*, Prentice-Hall, Englewood Cliffs, NJ, 1981.

[93] M. Boasson, The artistry of software architecture, *IEEE Software*, 12(6):13–16, November 1995.

[94] K. Beck, *Extreme Programming Explained: Embrace Change*, Addison-Wesley, New York, 2000.

[95] L. Bass, P. Clements, and R. Kazman, *Software Architecture in Practice*, Addison-Wesley, New York, 2007.

[96] V. R. Basili, R. Caldiera, and H. D. Rombach, *Goal Question Metric Paradigm, in Encyclopedia of Software Engineering*, vol. 1, John Wiley & Sons, New York, 1994.

[97] C. Alexander, S. Ishikawa, M. Silverstein, M. Jacobson, I. Fiksdahl-King, and S. Angel, *A Pattern Language: Towns, Buildings, Construction*, Oxford University Press, New York, 1977. DOI: 10.2307/1574526.

[98] E. Goldratt and J. Cox, *The Goal: A Process of Ongoing Improvement*, North River Press, Great Barrington, MA, 1984. DOI: 10.4324/9781315270456.

Author's Biography

SCOTT A. WHITMIRE

Scott A. Whitmire has been engineering and building applications of all sizes, from simple scripts to enterprise-scale applications that run multibillion dollar firms, for over 40 years. During that time, he has encountered many situations in which he had to know that the design was going to work before he was able to build and test it. Sometimes, this was due to a time crunch; sometimes, he just didn't want to spend the effort and have it fail. These situations led him to develop the practice of using well-known design criteria to evaluate design options during the design process. Many of these design criteria did not have suitable measures that would make them useful, so he invented them, along with a way to mathematically analyze the dynamic behavior of a design before there is code to test. The result is the software equivalent of structural analysis in civil engineering.

Mr. Whitmire holds a Bachelor's degree in Accounting and a Master of Software Engineering. He has written extensively on software engineering, system architecture, and business architecture, and has trained and mentored many engineers and architects. In addition to many presentations, articles, book chapters, and blog posts, he wrote *Object-Oriented Design Measurement* in 1997. His current day job is to design, build, and operate the software tools used by the Mathematical Neuro-Oncology Lab at the Mayo Clinic in Phoenix, Arizona, which looks and behaves like a $4M startup.

Printed in the United States
by Baker & Taylor Publisher Services